THE LIVING STAIRCASE

Arri Eisen

Emory University

David A. Westmoreland

United States Air Force Academy

KENDALL/HUNT PUBLISHING COMPANY

4050 Westmark Drive   Dubuque, Iowa 52002

All interior art by Rachel Park, except Figures 2.1, 2.2, 2.3, and 2.4 by Michael Kim.

Contents

Instructor's Preface

The Living Staircase is intended primarily as a companion text in introductory courses for *biology majors*.

We found that teaching biology as an interconnected whole is our biggest challenge in educating students, especially in large introductory classes. We noticed that what is most lacking in the hundreds of college freshmen we have taught is the ability to integrate the facts they have memorized into a bigger picture, to relate these facts to their lives. When we looked at our own classes, we saw that it was very difficult to establish a foundation of biological knowledge *and* constantly place this knowledge into the "big picture."

While we recognized this problem, we also realized students are already struggling to finish all the fact-based reading we assign them. One solution was to find a book that was informally written, and relatively short and non-intimidating that would integrate facts into the big picture using biologic examples and analogies from everyday experience. We were unable to find a book that filled this essential niche; so we wrote one.

We emphasize that the ability to integrate and understand fact and concept in biology is key to *all* students, but is especially important for biology majors. Educational materials that use strategies like ours to teach biology are too often "relegated" to non-majors courses. In our judgment, this is a mistake. Who should better appreciate the human relevance of biology and develop real-life connections to it than the people who are going to be our future doctors and researchers?

Our central metaphor is The Living Staircase. We take eight central themes of biology and explore each using The Staircase—the connected walkway of molecules, cells, organs, organ systems, organisms, and populations that compose life—as a guide. We illustrate the themes with examples from biology, providing enough background so that the examples we choose stand on their own, but leaving the details to a major text. The first chapter is as an introduction to The Living Staircase concept, and the last summarizes the interconnectedness of life with examples that unite all the themes.

Although *The Living Staircase* is designed with beginning biology majors in mind, the book would also be useful as a sole text in a non-majors course or as a companion text in a mid-level biology course. However you decide to use it, we wish you luck and encourage feedback concerning the books effectiveness.

We would especially like to thank the teachers that inspired us during our education, most notably Charles Lehman and Emma Bevacqua, and the many students who inspired us to write this book and helped us with insightful criticisms of the manuscript.

ESCALATORS

Theme

Biology is usually taught as a sequence of topics that increases in complexity, known as the Hierarchy of Biological Organization. At each level of the Hierarchy, novel features of life ("emergent properties") are found. The levels of the Hierarchy are interactive, so a change at a single level will affect the emergent properties of several others. To comprehend biology, it is essential to understand the interactive nature of the Hierarchy.

Examples

- The four levels of protein structure
- Exponential and logistic population growth
- Sickle-cell anemia

We are a moving bunch, the inhabitants of the twentieth century. Some history professor a hundred years from now will look back at us and take note of the ways we answered our desire to stay in motion with minimal exertion—cars, motorcycles, jets, high-speed trains, jet skis, speed boats—and will probably overlook the greatest innovation of all, the escalator. The evolution of this machine has been a remarkable adaptation for our couch-loving civilization. All the other means of transportation require some presence of mind to stay alive while using them; but the benevolent and trustworthy escalator delivers you to an intended destination while you daydream or sleep or read, regardless. It is nonjudgmental—you can ride passively or step along to speed things up.

It is no accident that I am fascinated by both escalators and biology. There are many parallels between the two. As a freshman in introductory biology, I remember learning "The Hierarchy of Biological Organization," which resembled a staircase but was actually an escalator in freeze-frame. Step One was molecules; Step Two, cells; Step Three, tissues; Step Four, organs; and so on up to the level of the biosphere (Figure 1.1; see foldout at back of book).

I saw **The Living Staircase** as being fixed in place—something to be memorized and stored away for an exam. But had I learned it as an escalator, with steps that interact, ebb and flow, I would have seen much more clearly into the science of biology. Escalators teach many lessons about life.

For example, they teach us about the development of coordination. The next time you are in a mall, take note. Watch a boy approaching an escalator—any two-year-old will do. He will move warily, pausing to eye the flattened steps as they emerge from some mysterious place below, and then he'll take a faltering step toward it. If he is too timid, Mom or Dad will urge him on—and catch him as he flails his arms in a wild attempt to regain balance. A year later, that kid will run and take a dive at it, and a parent will be there again, ready to catch him before he hits the grated edge of a rising step. A part of his brain called the cerebellum has been growing, and he is more confident about coordinating the complex muscle contractions required to board an escalator.

As a professor of biology, I sometimes see myself in the role of the parent (although I prefer the image of a big sister). My students in introductory biology are exploring a conceptual escalator, one that is not easily conquered, and I am there to help them through their faltering start. If I am successful, they will be running up and down the thing by the end of the year. Their cerebral hemispheres will have acquired a new level of thinking—one that opens their eyes to the marvels of life.

That is the purpose of this book, to explore the escalator. Not step by step, which is the approach taken by most huge introductory textbooks, but rather to go inside and explore the guts of the machine, the green flashes that most students only glimpse briefly between the steps. We hope to avoid presenting you with more information to be learned; rather, our goal is to rearrange and connect the things you have already learned, or are in the process of learning. For this reason, each chapter brings together examples from a variety of disciplines—examples that at first might seem entirely unrelated. We hope to reveal some of the underlying themes that run throughout The Living Staircase.

Emergent Properties

It is quite a challenge to connect the steps of The Living Staircase. Part of the difficulty lies in the fact that complexity grows with each step. At the lowest point, interplay between molecules gives rise to cells, where new features emerge—features that the biochemist, focused on step one, would never have the opportunity to observe. Then, cells coordinate and produce the next level, tissues, where complexity arises that is beyond the vision of the cell biologist. It is like that all the way up The Staircase, complexity building and spinning off new **emergent properties** which capture the attention of questionably sane people known as biologists.

Let's formalize the concept. *An emergent property is a feature that can be observed at one level of The Living Staircase, but not at lower levels.* Paradoxically, an emergent property arises because of the interactions that occur in the steps below. Imagine, for example, an emergent property that college students around the world thrive upon: music. One of my favorite groups, The Red Hot Chili Peppers, is composed of four people—Anthony Kiedis, John Frusciante, Chad Smith, and Flea. (All of these guys are out of the ordinary, but Flea is a bit farther removed than the rest.) If I listened to Flea play the bass guitar alone, it would probably sound good. I might enjoy listening to Chad Smith play drums by himself. But even if I spent a lot of time listening to every member of the group in isolation, I wouldn't be able to imagine the extraordinarily energetic music they produce together. Their music is an emergent property of the way they interact. Substitute one member of the group and you'll get a different sound.

In biology, understanding how an emergent property arises is essential knowledge. Consider the problem of a biochemist that has discovered a protein that protects fish from freezing in Antarctic temperatures. (Many of these fish live their entire lives at −2 °C; if the protein could be adapted for human tissues, organs might be stored for long periods instead of requiring immediate transplant.) How does the protein accomplish its job? The answer to this question, and many like it, has been delayed by a perplexing emergent property.

Any biochemist knows that the function of a protein is determined by its structure. When immersed in the aqueous environment of a cell, a protein coils and folds into an astoundingly complex shape—so complex that it has to be described at four levels: primary, secondary, tertiary, and quaternary (Figure 1.2).

The **primary structure** of a protein is simply its amino acid sequence. All proteins are made of repeating subunits called amino acids, just as all trains are made of repeating subunits called boxcars. The amino acids of a protein often interact to form various types of **secondary structure**. One of these is a spiral called the alpha helix. Above this level, the protein folds back on itself and amino acids interact again, creating larger loops called **tertiary structure**. Some proteins go further, creating **quaternary structure** when different amino acid chains combine.

It is hard to find a life-sized analogy to protein structure, but a rubber band can come close. Try this: take a rubber band by the two ends and twist it into a tight spiral. This is analogous to the secondary structure of a protein. If you keep twisting, the rubber band will curl into a higher-order loop, with the previously formed spirals inside. Tertiary structure in proteins is analogous to this. How could you mimic quaternary structure? By combining two or more rubber bands. In the rubber band, the spirals and loops (secondary and tertiary structure) are emergent properties that come from twisting the ends. In a protein, the secondary, tertiary, and quaternary structures arise from the way that amino acids interact. *Because these higher-order structures arise from interacting amino acids, they are emergent properties of the primary structure.*

If the primary structure of a protein is known, shouldn't it be possible to predict the other levels of structure? The task of deciphering primary structure was solved in the 1950s, and is now automated. It is fairly easy to do. The rest is not. The goal of predicting higher-order structure is known as the "protein-folding problem," and a problem it is. A protein goes through several intermediate shapes before assuming its mature form, so predicting the outcome is a daunting task. Imagine that you are given a jigsaw puzzle that requires each piece to be bent as you put it in place, then bent again to continue. Without a picture for a guide, how accurately could you predict the final outcome? (I will admit to having bent a few jigsaw pieces myself—out of frustration—but the final product was never anything recognizable. In a cell, the outcome of protein folding *has* to be right!) Solving the protein-folding problem is a major ongoing goal of biochemistry. Until this emergent property is understood, our biochemist will not be able to fully understand the antifreeze of the ice fish.

For a second example, let's move along The Staircase to the level of populations. Biologists who work at this level have identified two types of **population growth**. (The categorization is not perfect, but most species fall into one of the two classes.) Some species show **exponential growth** (Figure 1.3); typically, they are small

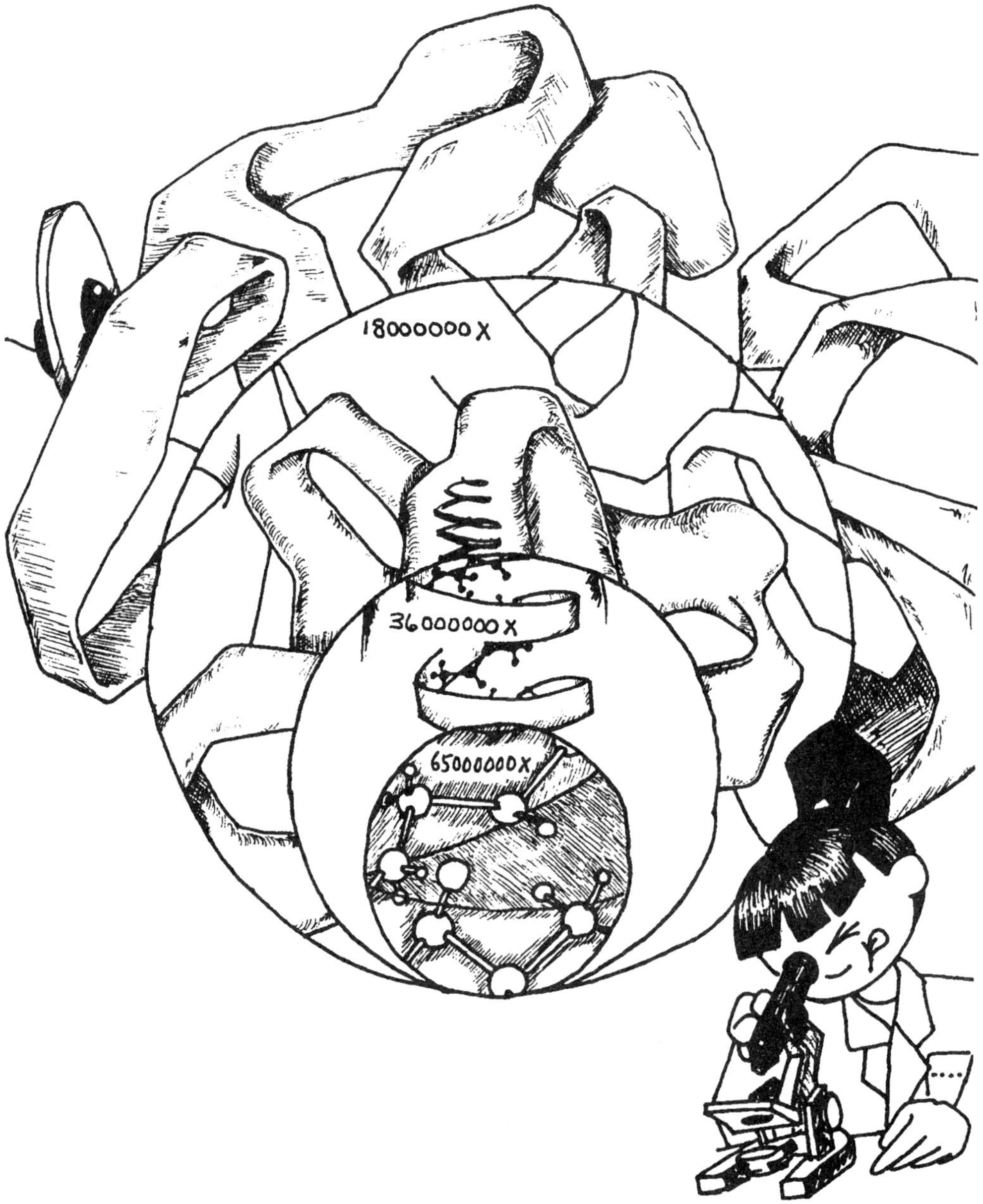

Figure 1.2 The shape of a protein determines its function. In a fluid environment, a protein folds into a shape that is described at four levels of complexity. Primary structure is the sequence of amino acids. Secondary structure arises from attractions between weak positive and negative charges in the amino acids. Tertiary structure is maintained by stronger chemical forces. Quaternary structure arises when amino acid chains combine as part of a single protein. Not all proteins have quaternary structure.

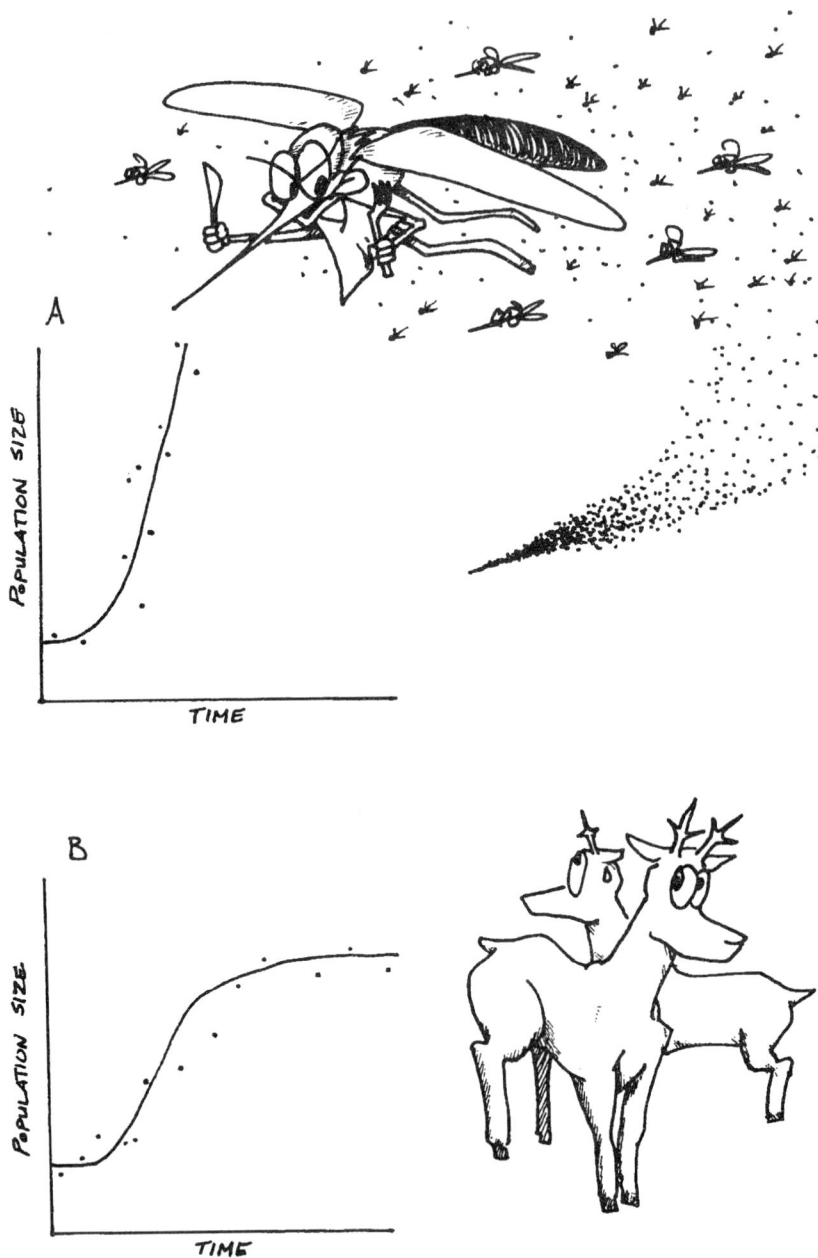

Figure 1.3 Two forms of population growth. (A) Exponential growth is typical of small organisms, like insects. Growth is extremely rapid, and continues as long as the climate is favorable. (B) Logistic growth is typical of larger organisms. The rate of growth slows down as the population density rises.

organisms that live in superabundant habitats. Mosquitoes are a good example—their populations virtually explode when environmental conditions are good. The second type of growth, shown by larger organisms, is **logistic**; the populations grow quickly at first, but the rate of growth slows as the population gets larger. Here we have two different emergent properties at Step 7 of The Living Staircase. How can they be explained? Because emergent properties arise from interactions taking place at a lower level, population ecologists have examined Step 6, the level of individual organisms.

It turns out that the two population growth patterns can be explained by the behavior of mating pairs. In populations that experience exponential growth, each mating pair attempts to produce as many offspring as possible. There is little or no parental care, so pairs can devote all of their reproductive energy to laying and fertilizing eggs. The result is extraordinarily fast population growth.

In contrast, logistic growth is the rule where parental care is essential. (Imagine a robin that lays its eggs but neglects to incubate them.) Parents expend some energy in producing and fertilizing eggs, but much more energy in caring for their offspring. As the population grows, it becomes harder to provide for the offspring because there are more adults competing for the same resources. The growth rate of the population slows because adults devote more energy to parental care and less to producing offspring in the first place.

Let's go back to the main point. An observation at the level of populations (Step 7) can be explained by the interactions of individuals (Step 6). This is the way explanations in biology generally work; we observe an emergent property at one level of The Living Staircase, and step down to discover the interactions which make that feature arise.

Communication in the Living Staircase

From time to time I come across an escalator that offers the opportunity to observe the whole cycle of human life. They are rare, but I know of at least two. One is in Washington DC; it rises from a subway station to the National Zoo. The other is in downtown Atlanta—and it has an attitude. The Peachtree Escalosaurus, as I have come to call it, is widely known for its abuse of commuters. It rises at a slow, excruciating pace from the deepest subway station to Peachtree Street. Only the sturdiest riders stand for the whole trip; most prefer to sit on the steps and wait out the time chatting with friends or reading romance novels. Some riders sleep. Others flirt and fall in love, marry, conceive children and have a small family before emerging at the top. (The order differs, often.)

On this very escalator I have observed a major characteristic of life—the flow of information.[1] One fall, I took eighteen freshmen through the underground circulatory system of Atlanta—basically to teach them how to find their way through the convoluted twists of arteries and veins known as MARTA (Metro Atlanta Rapid Transit Authority). We injected ourselves into the left brachial vein, otherwise known as Avondale Station, and whisked along to the heart (Five Points Station) where we got out to catch the carotid artery north. And finally, after all of this, we stood and admired the Peachtree Escalosaurus.

We stretched out to occupy a section some twenty steps long. I was the last to board, and looking up at the innocent little lambs under my charge, I felt that it was my duty to impart some sage advice. I tapped one of my students, Marcus, on the shoulder. He turned.

"Tell everybody to be sure they have a return ticket like this one."

He checked his, then tapped the next person, and the message began. Person by person, I watched the flow of information and the white flashes of subway tickets progress to the top.

I tapped Marcus. "Don't show your money, and don't give money to anybody."

He looked at me curiously. "Send it on," I said. He did, and I waited until I saw the person at the top nodding.

I tapped Marcus. "If you see a fight, don't try to break it up. Turn and go the other way."

His face betrayed exasperation. Marcus is tough—streetwise, a product of New York City. At eighteen, he has twice seen the barrel of a handgun used to settle disputes over pickup basketball. "The others," I explained, "are the ones that need to know this stuff." He passed the message on.

After a minute, a message flowed down the steps. "What do you think we are, idiots?" Before I could decide whether to respond honestly, the Escalosaurus regurgitated us onto the downtown streets, and my lambs dispersed in pursuit of adventure.

They found none that day. I sent eighteen hardy souls into the bowels of the city, and returned three hours later with an equal number. Different ones, perhaps, but still a personal record.

---

1. The transfer of information between organisms can be accomplished by sight, smell, hearing, or touch. Amazingly, all of these forms of communication share the same molecular basis. See Chapter 9 for full details.

Even the Peachtree Escalosaurus is dwarfed, however, by comparison to the big one, the less palpable escalator that I have ridden since birth and before. Indeed, I am on it now, and you are too. Here we go:

Level One: in my fingers, sodium ions drift across a membrane to the interior of a touch sensor; Level Two: the touch receptor depolarizes a nearby nerve cell, which jumps into action to produce electrical blips; Level Three: the blips rush up my arms and into my spinal cord, accelerated by insulating cells; Level Four: my brain evaluates the myriad of blips and formulates the appropriate response; Level Five: the nervous system coordinates to propel a series of impulses back down my arm; Level Six: a host of muscles contract and relax in orchestral synchrony, and the pen glides across the page. In the act of writing this paragraph, I have traversed more than half The Living Staircase. And in the act of reading it, you have done the same.

The point is, The Living Staircase is abuzz with activity. Information flows up and down, like messages to and from the freshmen on a field trip, and each step of The Staircase affects many others. Our current knowledge of biology enables us to trace the influence of one step through one or two other steps in most cases. But as the science progresses, we should become able to see the connections between all of the steps. There are a few areas in which we can do this now, and we will explore one now.

Sickle-Cell Anemia

As you read this, about a trillion red blood cells are rushing through the currents of your circulatory system, picking up oxygen from your lungs and delivering it to your tissues. Red blood cells are simple little things, living out their lives with a single function. When carrying oxygen, they are round and plump, like a fast-food hamburger, and when they have dropped off their oxygen, they collapse in the middle. (Rather like a cheeseburger that has been under the lights too long, but is still edible when you are cramming for a test at 3 AM.) That is, for most people they collapse in the middle. There is a second shape they can assume—the sickled conformation—that shows up in a small proportion of North Americans.

The two shapes are controlled by different versions (alleles) of a gene. Every person has two alleles, and the combination of alleles determines the type of red blood cells a person will have (Figure 1.4). Most people are homozygous normal, which a geneticist would call the wild type—they have two copies of the normal allele, so all of their red blood cells assume the collapsed-hamburger shape after releasing oxygen. A small number of people are homozygous for the other allele, and all of their red blood cells sickle. Then there are people that are heterozygotes, having one copy of each allele. In them, about half of the oxygen-depleted red blood cells sickle, and the rest assume the normal shape.

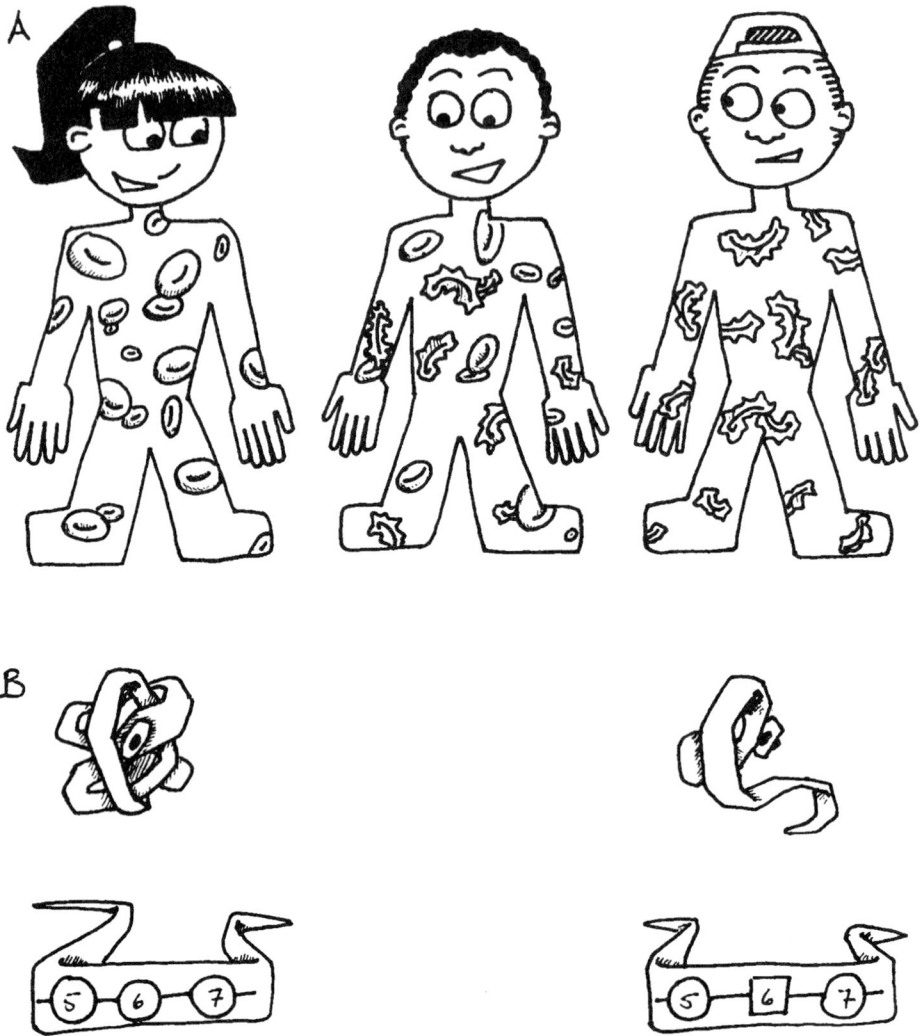

Figure 1.4 Sickled and normal cells in humans. (A) Person 1 is homozygous normal, or wild-type; the red blood cells are dimpled, but still round after they give off oxygen. Person 2 is a heterozygote, producing a combination of sickled and normal cells. This person is said to be a carrier, and the condition is called sickle-cell trait. Person 3 is a homozygote for sickled cells, and thus has sickle-cell disease. These cells become trapped in capillaries, impeding the flow of blood. (B) The cause of sickle-cell disease can ultimately be traced to a minor change in the structure of hemoglobin. (Left) Hemoglobin is composed of four polypeptide chains—two α chains and two ß chains. Each chain can bind to one molecule of O_2. (Right) In a person with normal hemoglobin, glutamine is the sixth amino acid of the ß chains. If valine is substituted, the hemoglobin molecules crystallize after giving off oxygen, causing the red blood cells to sickle.

Most interesting is this fact: homozygotes with sickled cells usually die from mild exertion before reaching ten years of age,[2] heterozygotes often feel intense pain after heavy exercise, and homozygotes with normal cells suffer nothing worse than standard exhaustion from exercising. We stand, then, at Step Six—the level of the individual—and want to explain this emergent property called sickle-cell disease. Take note as we explore the subject—sickle-cell disease is something that is well understood, and thus we can move freely up and down The Living Staircase. This is one subject in which the subdisciplines of biology are thoroughly interconnected.

The first question is, why is the pain (or death) of sickle-cell disease associated with physical exertion? The answer is fairly easy to come up with on your own, provided you realize that an emergent property of Step 6 is usually explained by the goings-on at Step 5. When a person is not exercising, most of the red blood cells are carrying oxygen, and thus are not in the life-threatening shape. When contracting muscles create a high demand for oxygen, however, more blood cells will release oxygen, and thus more will become sickled.

But why do sickled cells cause problems? Such cells have several characteristics that adversely affect organs and tissues. Sickled cells are fragile, having shorter life spans and thus reducing the ability of the blood to deliver oxygen. Sickled cells also tend to clump together and get stuck in the capillaries, cutting off oxygen to the tissues downstream. The clumped cells also increase the viscosity of the blood, demanding more work from the heart. Faced with a bleak future, suffocating tissues send alarm signals to the brain that are interpreted as pain.

Why do some red blood cells sickle, while others do not? If you are catching on to the basic idea of this chapter, you should realize that sickling is in itself an emergent property, one that occurs at the level of the cell. So, sickling probably arises from interactions among the molecules inside the cell. Indeed, it does; it is the result of the way that hemoglobin molecules interact. In fact, the alleles that determine whether a person's red blood cells will sickle do not code for "cell shape" directly; rather, they code for hemoglobin molecules inside the cells.

In a red blood cell, hemoglobin molecules are the workers that actually do the job of transporting oxygen. A typical cell contains about 250 million of these proteins, each of which is capable of carrying four molecules of oxygen at once. Hemoglobin has a complex structure (Figure 1.4). It is made up of four amino-acid chains—two are called the alpha chains, and the other two are the beta chains. The molecule as a whole contains 574 amino acids.

Now here's the critical point. In a person with sickle-cell anemia, the sixth amino acid of the beta chains is valine; in other people it is glutamine (Figure 1.4). Valine

2. New techniques in bone marrow transplant are extending the lives of these patients.

in position six causes hemoglobin molecules to crystallize when they lose oxygen, and as a result, they become bridges that span the cell and cause it to assume the sickled, life-threatening shape.

Amazing. One trivial change in the structure of hemoglobin—Step One—produces a series of emergent properties that determine life or death at Step Six. It is as if a domino is tipped over at the bottom of The Staircase and the effect is multiplied upwards until the structure shakes, on the verge of collapse.

Thus far we have **descended** the stairs in our search for answers. Now let's explore the upper end of The Staircase, the realm of the ecologist.

Upon learning of sickle-cell disease, an ecologist would wonder why it exists. It is inherited as a gene from one's parents. Since this gene leads to higher mortality, why is it present at all? We would expect people with sickle-cell disease to leave fewer descendants, since they don't live as long. Thus, over the centuries that humans have wandered earth, the gene should have disappeared from the human population. Right?

We can begin to answer this curiosity by looking at the geographical distribution of sickle-cell anemia (Figure 1.5). The disease is common in equatorial regions and is rare in the rest of the world. This is a hint that there is some advantage to having the gene in tropical ecosystems, an advantage that is lacking in other ecosystems. We step up, into the study of communities, to see what the advantage is.

Community ecology is the study of how species interact. In the case of sickle-cell anemia, the species of interest are humans, mosquitoes, and a blood parasite called **Plasmodium**, all of which are abundant in equatorial regions. Mosquitoes transfer the parasite into the human circulatory system, where **Plasmodium** infects red blood cells. The parasites mature and then emerge from the red blood cells as adults, producing fever and comas. The resulting disease, malaria, is severe enough to take the lives of more than two million people each year.

A sickled red blood cell is unsuitable for the parasite, and thus provides a degree of immunity. If 100% of a person's red blood cells sickle after giving up oxygen, the person has total immunity—but at the high cost of interfering with blood circulation. A person that has a 50% complement of sickled cells has an imperfect degree of immunity, but suffers much less from physical exercise. In equatorial regions, then, the person that carries a combination of sickled and normal cells has the highest survival, and leaves more descendants. Some 45% of equatorial Africans are of this type. In the United States, the sickle-cell allele is most common among African-Americans, who inherited the trait from their ancestors. It confers no adaptive advantage in North America where the malaria parasite doesn't exist, and conse-

Figure 1.5   The distributions of (A) sickle-cell disease and (B) Plasmodium, the cause of malaria, are closely intertwined. Carriers (heterozygotes) of sickle cell disease are resistant to the parasite, and thus have longer lifespans and have more offspring in equatorial regions. As humans have spread from Africa across the globe, the distribution of the sickle-cell allele has become widespread.

quently, the trait is slowly disappearing; currently, about 10% of African-Americans have a combination of sickling and normal cells.

Now, let's do a quick review of sickle-cell disease. At Step Nine, the ecosystem level, we make the observation that sickle-cell disease is most common in equatorial regions. Why? At Step Eight we find the answer—this is where humans, mosquitoes, and the **Plasmodium** parasite live as members of the same community. At Step Six, we observe that individuals with the sickle-cell trait suffer higher mortality from exercise, a fact that is explained by the way clumped cells (Step Two) block the circulatory system (Step Five), posing a challenge for the heart (Step Four) to work harder. Why do the deoxygenated red blood cells clump together? Because of a seemingly minor change in the structure of the hemoglobin molecule (Step One).

Theoretically, all of biology is connected this way, one emergent property integrated into another, and as researchers fill in the gaps, the steps of The Living Staircase will merge together like those of an escalator. In subsequent chapters, we hope to guide you on a quest to comprehend some of the underlying themes of biology, just like someone helped you onto your first escalator. By the end, perhaps you will have discovered themes for yourself.

? Q & A with Jeff & Jen

Jeff and Jen have a brief dialog at the end of each chapter. Since Jen is the author of this chapter, Jeff begins the questioning.

JEFF: I remember memorizing that Hierarchy of Organization. I guess I learned it like you did. My question is, why did I learn it at all? If all of biology is connected, why break it down into a hierarchy?

JEN: Biology is so complex that it has to be broken into pieces, and the hierarchy is a logical way to organize things. Check out the table of contents in almost any introductory biology book, and you'll see things laid out that way. Molecules first, cells next, and so on up to ecology. Even this book, where we're trying to show connections between the levels of the hierarchy, is divided into themes. But you have to remember that such divisions are artificial, a human-made system of organizing information. Life crisscrosses The Staircase without asking our permission.

JEFF: And that's where the connections come in. Good enough. Let's move on to emergent properties. I think I have the concept right—an emergent property arises from some type of biological interaction—and I understood your examples, but I can't think of any for myself.

JEN: Let's work on the definition first. It's vague—but then it's hard to define the term without a specific example. If you are standing at one level of The Living Staircase, say at the molecular level, you will see that the molecules are interacting. The result of this interaction will be observed at the next level, cells. So an emergent property is a feature observed at one level that is due to interactions at another level. Now, Jeff, think of an example.

JEFF: (pauses for a while). . . . I can't.

JEN: You're doing it right now. Thinking is an emergent property.

JEFF: What?

JEN: Right now, in your brain, there are several million neurons stimulating each other. The thoughts you have are an emergent property of neural interactions. This is what Francis Crick, the co-discoverer of the structure of DNA, is working on now.

JEFF: This is hard to believe. Something as fleeting as thought can be an emergent property? How can you prove this?

JEN: To be honest, you can't. That's why Crick refers to it as "the astonishing hypothesis." Instead of working directly with thought, he works with vision—which is a combination of several emergent properties. Cells in the retina interact to send messages to the brain—that's one emergent property. Then the brain integrates the messages to create a visual image of the scene; that's a second emergent property. And finally, another region of the brain makes us conscious of the visual image that has been created. That's a third one.

JEFF: That's just a lot of fancy language. You think these three emergent properties are independent. Isn't vision one continuous process?

JEN: Of course it's one process, but a process can be broken into parts. When a car is made, the engine is put together in one place and the body in another, right? Studies of people with brain damage indicate that vision is put together in pieces, too. One phenomenon called blindsight is truly amazing. People with this condition have a blind spot in part of the visual field. For example, a person may see just a black spot on the right-hand side of their field of view, even though their eyes are working fine.

JEFF: That's easy to explain. They're not getting the second emergent property because the occipital lobes of their brain are damaged. The visual image isn't being made.

JEN: Actually, the visual image is there—but they're not conscious of it. Experiments have proven this repeatedly. You take a person with blindsight and get them to keep their eyes fixed in place, then position two lights so that they fall in the blind region. Then you turn one on and ask "Which light is on, the green one or the red one?"

JEFF: And they say "How would I know? It's in my blind spot!"

JEN: Exactly. But if you ask them to guess anyway, they get it right virtually every time. Their brain is producing the visual image, but it's not elevating that image to the conscious level.

JEFF: Wow.

JEN: That was my reaction, too. Francis Crick applies the finding like this: A region of the brain can be damaged so that a person is not conscious of part of the visual field. This means that vision consciousness is an emergent property of interacting brain neurons. Maybe all of our consciousness is an emergent property. Maybe.

RELATED RESOURCES

The Astonishing Hypothesis: the Scientific Search for the Soul, F. Crick, Charles Scribner's Sons, New York, 1994.

Young blood: new life, R. Mestel, *New Scientist* 152 (November), 38–42, 1996.

Introduction to Protein Structure, C. Branden and J. Tooze, Garland Publishing, New York, 1991.

Water, water everywhere: subtly shaping protein structure and function, E. Pennisi, *Science News* 143 (February 20), 121–123, 1993.

The growing human population, N. Keyfitz, *Scientific American* (September), 119–126, 1989.

REINCARNATION

Theme

Life is characterized by a Division of Labor in which specific tasks are carried out by the components of an organized system. In many cases, each task is essential for life to be sustained.

Examples

- Mitochondria and ATP production
- Eukaryotic cell structure
- Volvox
- Cell specialization in humans
- Honeybee colonies
- Lichen structure

After finishing college, I took a year off before applying to graduate school, and it was not long before the parental source of funding ran dry. So I took a position as a reporter for a newspaper in Berkeley. It was my job to find interesting people to interview, and I was pretty good at it. At least, that was my opinion. The editor differed with me, however, and some of the best interviews I recorded did not see the satisfaction of print. The one I am about to recount was one of those.

My meeting with Ira Bloomfield was accidental, actually. I had parked my car just off campus to visit a bookstore, then returned to find a parking ticket and a dead battery. The ticket was cooperative—it shredded easily—but the battery was as obstinate as a spurned tattoo. After coasting down a mile of hills in a moronic attempt to jump-start the electronic ignition, I ended up on Telegraph Avenue with no slope to take advantage of. After midnight, on a Saturday. All college campuses have a neighborhood where eccentric people gather, but Telegraph Avenue—being an off-shoot of "Bezerkley"—gets the prize. I meandered through the unusual setting in search of a telephone, and it was then that I met Ira, local landmark and historian extraordinaire.

Mr. Bloomfield was peering through the window of a restaurant as I approached from behind. "Can you direct me to a public telephone?" I asked.

He turned abruptly and eyed me before responding. "I ain't seen you before. This your first time down here?" His appearance was not one that put me at ease—unshaven, bleary-eyed, needing a bath and a change of clothes. Ira was rather lax on hygiene in general, I observed while explaining my battery troubles.

"Well, I knew you weren't no regular Telegraph Avenue type," he correctly observed. "What do you do for a living?" I said that I studied and wrote about life.

"Life, huh? Well, what does a *kid* like you know about *that*? Why, you're barely out of diapers. I'll tell you about *life*, if you want to hear it."

I explained that I was behind schedule for my next story, and I really needed to run, and—

"Well, then," he said, grasping my elbow, "I'll make it short 'n' sweet. I know about life because I've lived it, you see. I've lived *all* of it. Up there at the college a guy can find professors that say they know everything—some of 'em even got Nobel Prizes—and they're just as fresh to the subject as a baby is. And do you know *why*? HUH?"

I shook my head and gently pulled out of his grasp.

"It's 'cause they don't have no memory, that's why! If they did, by gum, then they'd know a fact from fiction, and they'd quit them fancy jobs and live like I do. But it'll never happen. No way. *They* can't remember. But *I* can."

He drew close and sputtered the next words into my face.

"I been reincarnated a hundred times over, and I remember *all* of it, sister. I got the completest memory of anybody on earth. I been around—nobody else comes close. Some folks say they remember bein' Napoleon, or Cleopatra, or somebody famous like that. But I'm the only one—mark it down, now—the *only* one that recalls all the way back to when I was a prokaryote."

I was stunned.

"Surprised, huh? Didn't think I'd know about them things? Well, let me tell you, kid, I know *everything* you do, and a lot besides. That's what I meant about havin' a full memory. I remember it like it was yesterday. It was over three billion years ago, and still just as clear in my head as air is. Why, I can tell you anything you want to kn—excuse me a second."

Ira sauntered over to the curb and retrieved a half-smoked cigarette that had been discarded by a body builder with spiked hair. As he took a drag, the glow illuminated his leathery skin and gray whiskers.

I was caught between intrigue and apprehension. "You were a prokaryote?" I asked.

"Yeah. Three billion years ago. And let me tell you, times was tough. A cell had to make it alone. Back then,"—he paused here to take in another dose of nicotine—"nutrients was scarce. Had to swim all day just to find a few lousy amino acids, and carbohydrates were plenty rare, too. I tell you, kid, it was a *miserable* way to make a livin'. On account of there weren't no producers; this was all before photosynthesis, you know. All the molecules we depended on had to pop up on their own, like in that Miller and Urey experiment. You know that one?"

I said that I knew it well. Miller and Urey were the first researchers to experimentally test a theory on the origin of biological molecules. They set up an apparatus that simulated the early environment on earth, and proved that amino acids and other molecules could arise autonomously, without being made by cells (Figure 2.1).

"So you do know somethin', I guess. But let me ask you this: What's the point of spending all that money and time when they could'a just as well asked me? I knew it already, and I was at Chicago back in '53 when they done that experiment!"

Figure 2.1   Where did the first cells get their nutrients? (A) According to theory, nutrients like glucose and amino acids formed spontaneously when the earth was young. Such nutrients were scarce, but sufficient to keep life going. (B) Miller and Urey tested this hypothesis by setting up an apparatus that simulated the earth's environment at 3 billion years ago. They showed that nutrients could form spontaneously.

I shrugged and redirected the inquiry. "What was it like back then—not in '53, but back when you were a prokaryote?"

"It was hotter 'n' a heat wave in Havana—up around the boiling point all day, every day—really, I ain't tellin' a stretcher. But that wasn't no problem. We was adapted for it. Nah, it was the lack of nutrients that done most of us in. I hung on for a day, but just wasted away after that.

He threw away the cigarette, now with the cotton filter burning, and at that point in the evening I admit that I had mixed feelings about Mr. Bloomfield. His sanity was certainly in question. However, his knowledge was right on target. The first cells did arise around three billion years ago, and they had to scavenge for scarce nutrients until photosynthesis evolved. And it was hot at the time—this was when the earth was young and was still cooling down. Bloomfield's combination of knowledge and storytelling was sublime, and I realized that I was onto something worth pursuing. I offered a meal and he accepted, and we stepped into the deli.

"What did you do after you died?" I asked.

"I did what we all do. I went back to the ether and got in line for another life."

"No, I meant, what was your next life?"

He peered at me over the top of the menu. "Where does it say 'coffee' on here?"

I pointed to the line on the menu. "Just tell the waiter what you want." While he ordered I jotted down the date, time, and location, like a good cub reporter.

"I waited a billion years and then come back pretty much the same, but in a lot better place. Let me tell you, sis, a billion years is a long wait. But you don't have no control over that. You got a number, and you got to set there until it comes up. Some of the other spirits got whisked away after just a thousand years. Just goes to show there ain't nothin' fair on heaven or earth, and you can forget whinin' about it. You listenin'?" He leaned back in his chair.

"Did you come back as another bacterium?" I prompted.

"Well, I was a bacterium of sorts. But in a great situation. Resources? No problem. Why, I could 'a taken in amino acids all day long and never run out. I was an insider, see."

"An insider," I said as I wrote.

"Yep. Hmm, this is good coffee. Haven't had a hot and more 'n' half full cup in a week or so."

"My pleasure. But please, go on."

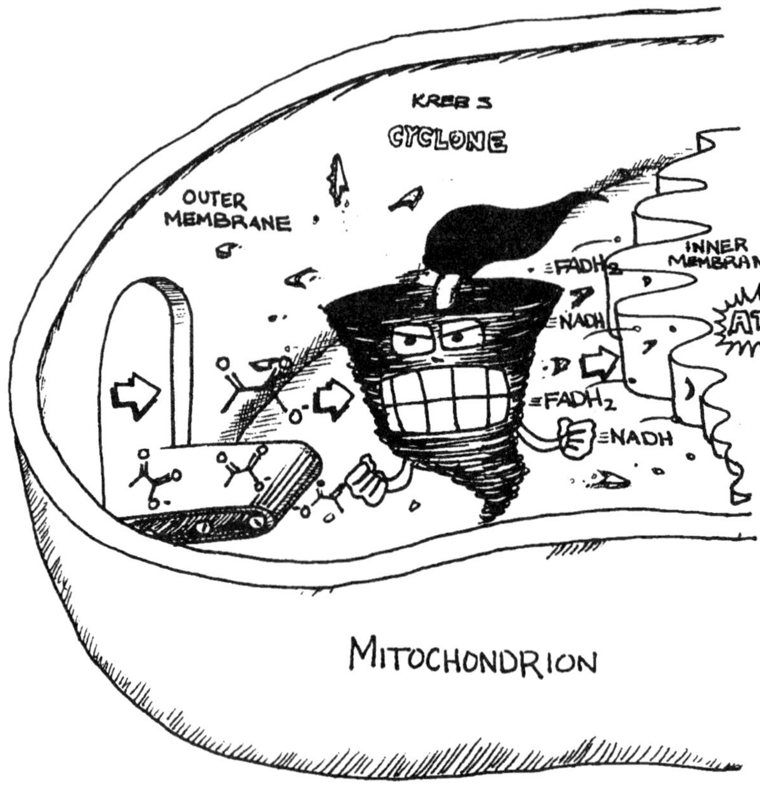

Figure 2.2 A mitochondrion provides an example of the Division of Labor at the intracellular level. The outer membrane transports pyruvic acid to the interior, where it is broken down by the Krebs cycle. Then, the products of the Krebs cycle are delivered to the inner membrane, where they are used to make ATP.

"Oh, yeah. It was sweet in that cell. Luxurious, even. All I had to do was pick up the nutrients I needed out of the cytoplasm—mainly pyruvic acid, you know—and run 'em through the Krebs cycle[1] and the electron transport chain[2] , and I'd make all the ATP I needed (Figure 2.2). Extra, really. I used to give lots of ATP away—to the other organelles, I mean."

---

1. Details of the Krebs cycle are given in Chapter 3.
2. The electron transport system is described in Chapter 8, which covers the principles of thermodynamics.

It took me a moment to put it all together. He was something inside a cell, something that used pyruvic acid to make ATP. "You were a **mitochondrion**?!"

"Well of course! What else? Didn't you say you was educated in biology?"

"I have a degree in biology," I said indignantly, "but as a reporter I have to make sure that I have the facts straight."

Ira guffawed and shook his long clumpy hair so wildly that people stopped and looked through the window at us. "Since when does a *reporter*—"

I saw that one coming and cut him off. "Right. I got it. A mitochondrion," I scribbled aloud, "in one of the early eukaryotic cells."

"Right on the mark, kid."

A compliment! At least it sounded like one. Our food arrived and Ira began to eat, and it became clear that the interview would have to be put on hold. But then he surprised me by being able to talk around a full mouth. "It was when I was a mitochondrion," he said, "that I first learned about *The Principle*." And he leaned forward and winked at me when he said the last part, as if revealing a pearl of wisdom. "You know what that is?"

"Well of course," I said. "Everyone who has studied life to any extent is familiar with it."

"Good. Maybe college ain't a complete waste." He drained a second cup of coffee in one gulp and ordered a third. I sipped at mine and came up with a strategy.

"Mr. Bloomfield," I said, "you are a remarkable man. I was lucky to run into you this evening."

He looked surprised and moderately pleased, and seeing that I was off to a good start, I gave him another dose. "And there's another thing I've noticed about you. Your eloquence. You express yourself very well. My readers will enjoy this piece."

"Well, now, that's a nice thing to say," he said. "But I ain't tellin' it for their sake, you know."

"Still, this should be shared with others," I said. "And I'll need to help the readers out somewhat; most of them are not biologists, you see, and, um, consequently, they could not be expected to be familiar with the subject."

I paused, then, and waited, but Ira gave me a blank expression. Silence divided us.

"In particular, with regard to The Principle," I added.

Still he made no move to clarify. Until then, I had considered myself an expert at digging out information by subtle means. The silence lasted an hour, it seemed, before I took another swing at it.

"I would go ahead and add the information myself, you know, but I don't think I could explain it as well as you could. I fear that I might, um, obscure it—The Principle, I mean—by adding too much technical detail."

Bloomfield smiled and chuckled to himself for a moment. "Oh, that. Well, the way you handle it, kid, is just to call it the **Division of Labor**. People see that every day—heck, they think it was invented by humans, but it was goin' on way before there were any two-legged half-wits on the planet."

"Right," I said. "No problem. I need to get across the idea that a mitochondrion helps to, uh, let me see—divide up the work that has to be done in a a cell."

"That's good for a start, but it ain't enough. 'Course, it's so obvious in everyday life that nobody'd think nothin' of it. Why, just look around us. This here waiter takes our order; the cook in the back whips it up; we eat it and pay for it—sometimes more ways than one. But what the average person don't get is that The Principle goes on everywhere. Everywhere you turn a microscope, and most of the places you don't, there it is.

"Back when I was a mitochondrion, I could look inside myself and see it clear as a bell. I had two membranes, one wrapped all around the other, and they did completely different jobs. The outside one brought in pyruvic acid to get broke down in the middle—that's where the Krebs cycle was runnin', and then the inside membrane took the stuff from the Krebs cycle and made it into ATP (Figure 2.2).

"And then I could look outside myself—all around me—and see The Principle again. All us organelles worked together, more or less, to keep the cell goin'. Not just us mitochondria, but everybody. The nucleus sent out messages—messenger RNA, you know—about what proteins needed to be made, then the ribosomes'd read the message and make the proteins, and then the Golgi apparatus or the endoplasmic reticulum'd shape up the proteins to get 'em ready to work (Figure 2.3). Most of all, I liked it because I felt important. The whole bunch, all of 'em in the cell there together, depended on me to keep up the ATP supply. 'Course, I depended on them, too. I had to have some of them proteins that was bein' made. We weren't independent like I was in the first life—none of us was. We were, uh, what's that word?"

"Symbiotic. You lived in close contact and interacted extensively."

Figure 2.3   The eukaryotic cell is a model example of the Division of Labor. Each organelle carries out specific functions essential to the survival of the cell as a whole. The nucleus is the central organizing center; it sends out chemical messages (messenger RNA) telling which proteins should be made; ribosomes read the messages and make the proteins; the rough endoplasmic reticulum and Golgi body modify and package the proteins. Other organelles, like lysosomes, contain enzymes that carry out other functions.

"Yeah. It was somethin' to see. Communication? My gosh, you should 'a' seen all the chemical messages flyin' around.[3] To and from the nucleus, between me and the other organelles, even to the outside. That's what made it all work. Can't have a division of labor without lots of communication. Why, I remember sometimes the cytoplasm'd make more pyruvic acid than me and the other mitochondria could handle, and we'd send out a chemical message that'd slow 'em down for a while. Picture of organization, that was us."

He paused a minute to let me catch up on the notes, and then he gunned the engine and took off again. "You see, sis, when you bring a bunch of living things together to live in the same place, they'll start dividin' up the jobs, and pretty soon they depend on each other. One group quits, they all crash 'n' burn. When I died as a mitochondrion, why, that cell would've been a goner if there hadn't been a bunch more like me in there."

"How did you go that time?"

"Just wore out after a while. Got old and petered out."

"So then you started a third life. Were you a complete cell that time?" I asked.

"Hey, you're catchin' on. Got to be an amoeba on that run, and back then, an amoeba was one of most complex critters. I remember oozin' and slimin' along, searchin' out bacteria and taken 'em in for food. Felt kind of bad about that, bein' as I'd seen the bacteria's look at things, but that's the way life is, ain't it? If you ain't a photosynthesizer, generally, you got to eat other things."

"How long did you live?"

"About a week. Don't seem like much now, but I tell you, sis, you can learn a lot in seven days. Every minute while I was slimin' around I saw The Principle in action. Not around me, see, on account of I was a lonely cell, but *inside* me. I had all the organelles of a regular eukaryote and they just did the finest job ever. Nucleus was fine-tuned as it could be—kept everything coordinated."

"But you had already learned that, back as a mitochondrion. How was being an amoeba any different?"

"It was the perspective. I'd seen organelles before—heck, I'd been one—but this was different. I saw the organelles kinda like from above—I could look inside myself and see all the organelles at once. There ain't nothin' that clears up an idea for you like seein' it from different angles."

"So what happened?"

---

3. Virtually all forms of communication have a chemical basis, as discussed in Chapter 9.

Ira looked as if he didn't understand the question.

"How did you die?" I asked.

"Didn't. I just growed up to full size and split in half. See, when that happens, you got to leave and turn things over to two new spirits. That's the way it works—ever' life you learn a little more about somethin' and then you get to the next level. That's the scheme of things up there." He pointed to the ceiling and I glanced upward in hesitant agreement.

"How many lives do you have left?"

"How the heck would I know *that*? I *live* life, kid, I don't make the decisions!"

I felt my face turn red, and he softened his tone. "Sorry; suppose I'm a little sensitive about bein' set back. But anyways, like I said, I slimed along with all them little organelles inside, and I figured I'd gone as far as I could. Thought I'd seen The Principle through 'n' through. Thought I had it mastered. And then I come back as a **Volvox**" (Figure 2.4).

"You were a *Volvox*?"

"You know what one of them is?"

I nodded. "It's a colony of photosynthetic cells. A primitive form of multicellular life." And in saying that, I recognized the trend in Mr. Bloomfield's reincarnations. As a mitochondrion, he had spanned the bridge between Steps 1 and 2 of The Living Staircase; as a eukaryotic cell he had stood firmly at Step 2, and as a *Volvox,* he was exploring Step 3.

"Well, that surprises me. I didn't figure you'd know that one. It was around 800 million years ago. Boom! All of a sudden everything's different. I'm floatin' around there, looking myself over, and I see The Principle two-layered."

"You'd better explain that one to me," I said. Ira seemed pleased that I was bowing to his superior knowledge.

"See now, I was maybe a thousand cells, all together, but they was specialized. Some did the job of reproducin' and others moved us around. The cells on the outside had flagella, and they could beat 'em and I'd float off this way, and then them on the other side'd beat theirs and I'd move somewheres else. That was The Principle at the level I was at. But at the same time, I knew that inside each one of them cells that made *me* up, there was all the organelles I had back when I was an amoeba. So there it was at another level, simultaneous."

Ira wiped his mouth on the corner of the tablecloth and belched politely, then brushed his hand across his whiskers to remove the fallen remains of the meal.

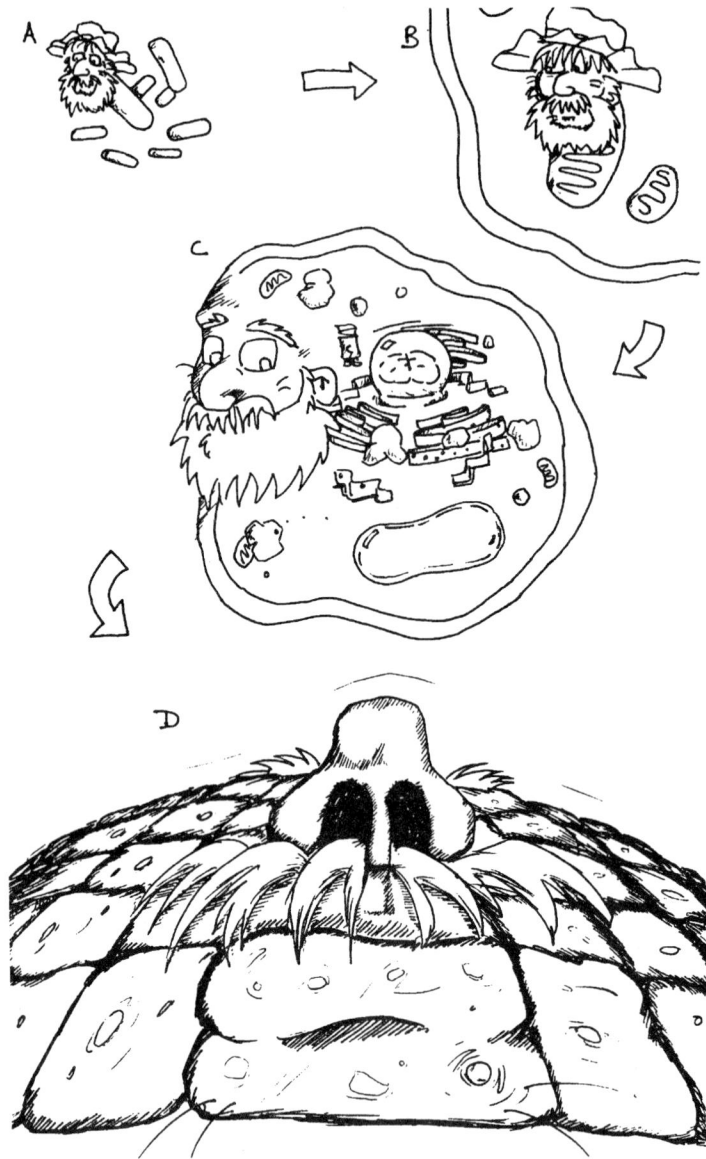

Figure 2.4   Ira's first three lives show the Division of Labor at three levels of the Living Staircase. (A) As a bacterium, Ira did not notice The Principle, probably because his life was so brief. However, it was present at the intracellular level. The bacterial cell contains ribosomes that make proteins, and a nucleoid that directs which proteins are to be made. (B) In his second life, as a mitochondrion, Ira noticed The Principle at the intracellular level. (C) Then he saw The Principle in action from a different perspective, at the level of the cell. (D) As a *Volvox*, Ira observed the division of labor among a thousand cells simultaneously—a prototype of a tissue.

I hardly noticed. I was so absorbed in his statement that I was dazed, staring into empty space with a hurricane in my head. This was a revelation. What Ira had said was just the beginning. Prokaryotic cells came together to make the first eukaryotes, right? And then the eukaryotic cells grouped to make multicellular creatures like the *Volvox*, and the Division of Labor mushroomed from there—the cells in multicellular creatures specializing into various tissues, tissues interacting to form organs, organs to form systems, systems interacting to make an individual. In a way of thinking, I was just a complex symbiosis of prokaryotic cells!

"Can I get a brownie?" Ira said, and my thoughts came back to earth.

"Yes," I said. "Sure."

"And more coffee."

I motioned to the waiter and got back on track with the interview. "What other important points about The Principle should my readers know?"

Ira turned away suddenly and sneezed so hard that the table shook, then wiped his nose on his sleeve and said "They ought to know how come things do it," he said. "Divide up the work, I mean. It's for efficiency, that's how come. Them that's more efficient live longer and reproduce more."

I nodded, and noticed that the people at the next table were leaving without having finished a meal. "The division of labor leads to higher efficiency," I summarized as I wrote.

Ira nodded. "Now you tell 'em, kid, and tell 'em straight—you can see The Principle everywhere, not just at the cell level. Look at your hand right now, the left one—different muscles are twitchin' this way and that way, just at the right time to keep the pen movin'. That there's The Principle at the organ level. And what keeps them muscles organized? Nerve impulses from the brain, and oxygen and glucose from circulation—that's the thing in operation at the system level. Different systems providin' everything necessary to keep the muscles workin'. And look at yourself. A human. You got, what?—64 trillion cells?—and every last one of 'em's basically the same. Same DNA. Let that'n soak in for a minute—same DNA in every one of them cells, on account of they all came from the one cell you started out as, and what've they done? Divided up everything! Reproduction by one bunch, sense of touch by another, sight by a bunch more, and so on forever. And when one set quits, all the rest suffer" (Figure 2.5).

Another astounding thought. All of my cells are genetically identical, but functionally specialized. I let it soak in before continuing. "How many lives have you had?"

Figure 2.5 A human develops from a single cell, the zygote, into a complex of 64 trillion cells. All of these cells contain the same genes, but any particular cell makes use of only a small proportion of them. This is how cells specialize for particular functions.

"Oh, let me see." He began to enumerate on his fingers. I been a layer of smooth muscle in the gut of a cockroach, a semilunar valve in the right ventricle of a naked mole rat,—"

"Hold on, now," I said. "You can be those things? Just pieces of animals?"

"Sure. Plants, too. I was an an ovule in a lily flower once."

"Weird." I was thinking aloud.

"Well how else you goin' to master The Principle? You got to see it from every perspective, understand. You got to be an organelle, then a cell, a tissue, an organ, and *all* that."

"Is this your first time to be a human?"

"Yep. And *last*." Ira looked down on himself disgustedly. "Anybody who knows the big picture wouldn't brag about bein' a human. You can take *that* to the bank. Hear?"

I thought it wise to shift to a lighter topic. "What would you say is the most interesting life you've had?"

"Easy. It was the time I was a **honeybee**. Talk about order and purpose! I tell you, us humans ain't got nothin' like it. Look out the window at them people millin' around out there—punkers, dopeheads, researchers, students, plumbers, paper-shufflers galore, lawyers—you think they got a sense of purpose? They might tell you they do, but they don't even come close. Their work and resources is divided up, and every type is important to make the system work, 'cept for the lawyers, maybe. They all think they're gettin' along on their own. Humans are ignorant—rich in ignorance! It's their plainest feature, they say up there in the ether.

"But in honeybees, The Principle is all you see. The colony I was in had somewheres around 30,000 bees, and was all closed around the queen. All honeybee colonies are like that, on account of she's the only one that can reproduce. Yeah, that was The Principle nailed down tight. Every last job is covered—some of the bees spend all day groomin' her, others carry off her eggs and put 'em in growth chambers, and others feed the little ones when they hatch out.[4] Why, there's bees with wax glands that add onto the hive, bees that guard the entrance, and other bees—most of 'em, actually, that fly out and find flowers and bring back food for everybody else. Kid, when a colony's really humpin' along, there's fifteen hundred bees dyin' ever' day, in service to Her Majesty, you see, but she cranks out enough eggs to make up for 'em, and then some! I know that for a fact, 'cause I made a point of checkin' it out back then."

---

4. This division of labor also demonstrates behavioral diversity, which is discussed in Chapter 7.

"So this was your experience at the population level. When did it occur?"

"It was the life before this one here."

"What was your job at the hive?"

"I did all of 'em. See, when a bee's full grown, she comes chargin' out of her chamber and goes through all them jobs I was talkin' about. She works the nursery first, then adds wax onto the hive, and so on. And sacrifice! Why, she'll do anything she has to. I say 'she' 'cause most of 'em are females, you know. There's a few males, maybe a hundred, but they're no account. Don't do no work. Just wait around for a shot at matin'.

"But what was I on to? Oh yeah, the sacrifice a bee makes. Like I said, I worked all the jobs, and ended up at the end as a forager. Talk about work! Sis, you don't know the meanin' of the word. There ain't a human on earth that's ever worked hard as a bee. Flyin' out at the crack of dawn and bringin' back pollen and nectar, and waggle dancin' to tell all your sisters where the flowers are at—again and again like that, all day long until the sun sets."

"Waggle dancing?" I asked.

"It's the way a bee tells the other bees where the nectar is," he explained. "If you was to find a patch of flowers that was in line with the sun, you'd go back into the hive and dance straight up. Straight up means 'fly toward the sun and you'll find it' to the other bees. And if it was straight away from the sun, you'd dance straight down. Like that. Get the picture?"

I nodded. "It's a form of communication.[5] Keeps the foragers organized."

"Yep. Anyway, I worked my wings off back when I was a forager. I ain't complainin', and I didn't back then, neither. I liked it, even though I didn't see no purpose in it. Anyway, my wings finally wore down to nubs, and then I took to walkin' out to find flowers. Finally ran out of steam and rolled over dead. It ain't braggin'—it's the truth. And it ain't no different than what any of the others was doin', or what bees is still doin'."

I interjected when he paused, "How do you like being a human? I know you don't think we live by The Principle like the bees do, but other than that, what do you think?"

"You got me wrong there, kid. You live by The Principle just like everything else—it's just that you don't think about it. But about bein' human—I don't care for it. Not at all. The only part I like is seein' all these people think they're so high and mighty, and just the grandest form of life on the planet—and knowin' that when they

---

5. Visual communication, seemingly unique, shares several fundamental similarities with other forms of communication. Chapter 9 covers this topic.

get back up there to the lottery house they're goin' to try their level best to keep it a secret that they was a **Homo sapiens**." Ira chuckled at the thought.

"Why?" I asked. "Why would you be embarrassed about having been a human?"

"You'll see soon enough," he said, and a shiver ran through me. But then he reconsidered. "Well, since you've been so nice to me, I'll let you in on it. Bein' human, see, is kind of like bein' held back in grade school. Some folks is unlucky enough to draw it by chance, but most folks flunked a celestial exam somewhere along the way."

I wondered what grave error I had made in a previous life. "So how did you end up here?" I asked.

Ira looked to the left and to the right, then leaned over the table and spoke quietly. "To tell you the truth, I blew it when I was a bee."

My eyebrows rose, imploring him to explain.

Ira hesitated, motioned for more coffee, and began. "There was a week in there where I was a guard. Our job—there was a bunch of us, see—was to hang around the openin' and make sure that only our sisters got in."

"Your sisters?"

"Yeah. All the bees in a hive come from the queen, so they all have the same mother, and some of us had the same father. So all 30,000 of us were at least half sisters. See?"

I nodded.

"Well, now, sometimes a bee from another colony 'd get lost, or try to sneak in, and we had to chase 'er off. Easy enough. But on top of that, we had to go after every other animal that was a threat, too, and they was usually a lot bigger than we was, so we was supposed to use our stingers in case of that.

"Well, one day I'm sittin' out there enjoyin' the fine weather, and here comes this boy wanderin' around lookin' for trouble with a BB gun. We didn't know what one of them was, of course, bein' bees, but we straightened up and watched him like a hawk anyways. He popped off at a couple of birds—seemed like that was mainly what he was after, but then he saw us and took a shot. And hit, too—that pellet ran right through the hive and out the other side. Didn't do no damage to speak of, 'cept to kill a couple of workers and a larva or two, but still, we couldn't allow such things. In a split second the other guards was divin' at him, and I was bringin' up the rear. Hollerin'? You never heard so much noise in your life. He flung that gun down and run like a scalded dog, all the way swingin' his arms around his head like a faith healer. I buzzed him a couple of times, just for fun, you know, then flew back and laughed with the others. We was excited, I tell you—doin' high-fives with our antennas and scramblin' all over each other and wrestlin', and then I saw somethin' to sober me up right quick."

"The boy was suffering?"

"No, I figured he deserved what he got. What I saw was some of the guards that stung him flyin' in. What a sight! They was barely able to land, and hurt terrible, too. Couple of 'em quivered a little and just fell off the edge, stone dead. The rest crawled around maybe an hour, buzzin' with the pain before they went on. That's when I found out about sacrifice, see. When you sting somethin', the stinger rips out and goes with the kid, or whatever animal it is. It keeps on pumpin' out the poison for a good while, too, which is a slick adaptation, but the flip side is, that stinger takes a good part of a bee's guts with it."

"So, defending the hive is an act of suicide," I suggested.

"More or less. And the guards ain't the only ones that'll do it. The workers will too, if they're called out for reinforcements. In fact, every last bee in the hive dies sooner or later, in service to the queen."

"What about the males?," I asked. "I thought you said they don't do any work."

"They don't. But when they do finally mate, the same thing happens. Their guts get ripped out."

"Wow."

We sat in silence for a second before I realized that we had taken a sidetrack. I reminded Ira that he was going to explain his mistake.

"Oh, yeah. Well, after I saw what happens when you sting somethin', I decided to just keep my stinger to myself. I flew out and buzzed and raised cain whenever it was called for, and nobody ever knew the difference. But I never stung nothin'. See, I just couldn't see why I should sacrifice myself for the queen like that. I didn't mind workin' my wings off for her—I liked the work—but why cash out when you're havin' a good time? So I just waited 'till my next post came up, and went on to bein' a forager."

"I can see your point," I said. "I probably would have done the same thing."

"Maybe you did, sis. Maybe that's why you're sittin' here with me. But listen—just in case you *did* mess up the same way—I've had a lot of time to think this thing over, and I see where I went wrong. That's the thing about bein' a human—why it's a good demotion. You ain't got much in the way of prestige, once you get back up to the ether, but what you *can* do is think. Big cerebral hemispheres, you know. Back as a bee I figured my whole life was purely set to serve the queen. But it wasn't. Since then I've figured out that I was servin' myself, on account of she was a relative. Every time she laid an egg, there was a part of me in it, see, because me and her had some of the same genes, and she put some of them same genes in every last egg she laid. So anything I could do to keep her goin'—including suicide—was in my *own* interest. That's the way it always works. The Principle shows up all over the place, but only because the division of labor's got some benefit."

I recognized this as the concept of **kin fitness** (Figure 2.6). An organism can reproduce directly, by having offspring, or indirectly when its relatives have offspring. A good example is my niece. Because my brother and I have some genes in common, and he passed a portion of these to her, my niece is carrying some of my genes. In effect, I have been partially reproduced. I can also reproduce on my own, of course. But since a honeybee can only reproduce through her kin, the queen, it is in her interest to sacrifice anything to keep the queen alive.

"You got demoted for that?" I asked. "It seems like a trivial mistake."

"It's a big deal, kid, not to appreciate the basics of a thing like The Principle that's slappin' you in the face everywheres you turn. I won't make that mistake again, I can tell you right now. I'm ready for my next shot."

"Are you saying you want to go ahead and die?"

"I'd say I ain't afraid of it, anyways. There's a good side and a bad side to it, you see. Death ain't fun, but you got a new adventure comin' up once it's done. The waitin', on the other hand, has got to be about the terriblest torture a soul can stand. You never know how long you're goin' to have to sit before your number comes up again. I been whisked off in minutes, and had to sit near on a billion years at other times. All that waitin', and waitin', and waitin'—and you're all huddled together in the dark, maybe a trillion of you, all with your fingers crossed at once.

"You might think a bunch of folks that've lived a hundred lives apiece 'd have a lot to talk about, but after a millennium of hearin' 'Well, back when I was a six-toed cat . . .' and 'You think that's rough? You oughtta try bein' a dung beetle . . .' and 'Heart valve? Heck—I was a urinary bladder in a freshwater frog, and I remember bein' stretched out so far . . . ,' you'd just as soon a-died for good. But I'll face it again, sure as shootin', and this time I'll be happy just to be back on track. Progressin' to higher levels, I mean."

"What do you want to be next time?"

"I don't get to choose, of course, but I've got my fingers crossed in hopes of becomin' a **lichen**."

Figure 2.6 The concept of kin fitness. Siblings have some genes in common, because they have inherited their genes from the same parents. When one sibling reproduces, some of the genes are sent into the next generation. In other words, a part of her has been reproduced. The "jeans" symbolize how the genes mix from generation to generation.

It struck me as a strange goal. I would have thought a lichen to be a step backwards. "Why do you want to be a lichen?"

"It's the next logical level to somebody that's been studyin' The Principle. See, I was part of it at the level of the cell, back when I was a mitochondrion; I lived it at the tissue level, as that smooth muscle, then I rose through the other levels until I got up to studyin' at the population level."

"That's when you were a part of the bee colony."

"Right, and I ran into that little problem. So what's next? You tell me."

"It would be Step 8, the community level," I said, referring back to the steps of The Living Staircase I had learned as a freshman. "Interactions between two or more species."

"That's where a lichen comes in, sis (Figure 2.7). It's two species, you know—a fungus and an alga that turn into a whole new type of life by livin' together. One part of you, the fungus, sets up a decent place for the other part to live, and the alga makes food to keep everything goin'. It's back to the same situation I had as a mitochondrion, 'cept at another level. I figure I'll have gone about as far as I can when I get up to that point. But then, I thought that when I was an amoeba, too."

I paid the bill, and we stepped back onto Telegraph Avenue. He shook my hand and said "I like you, sis, and you know why? You didn't try to 'help' me. There's people runnin' up and down here all day and night, thinkin' they got somethin' I'm just dyin' for, and I could flush 'em all, sometimes. I figured you were goin' to be one of those types—that's why I was a little tough at first. I just like to think, that's all. I don't want nothin' from nobody. I had a good time talkin' to you."

I said I was glad to hear that we both got something out of the conversation, and Ira gave me a little smile and said "One more thing, kid. Don't try to fake it when you don't know somethin'. It never fools nobody. Bein' ignorant ain't nothin' to be embarrassed about—it's only if you stay that way that you oughtta to be ashamed." He winked, then turned and walked down a dimly lit side street. My final view of Ira had him peering into a dumpster.

While I sat in my car waiting for R & W Towing to bring rescue, I scanned my notes of our talk and marveled at how much I had learned from a single conversation with a man that I would have passed by—even avoided—on any other day. And I wondered how much knowledge and wisdom I am surrounded by, but unaware of, every day.

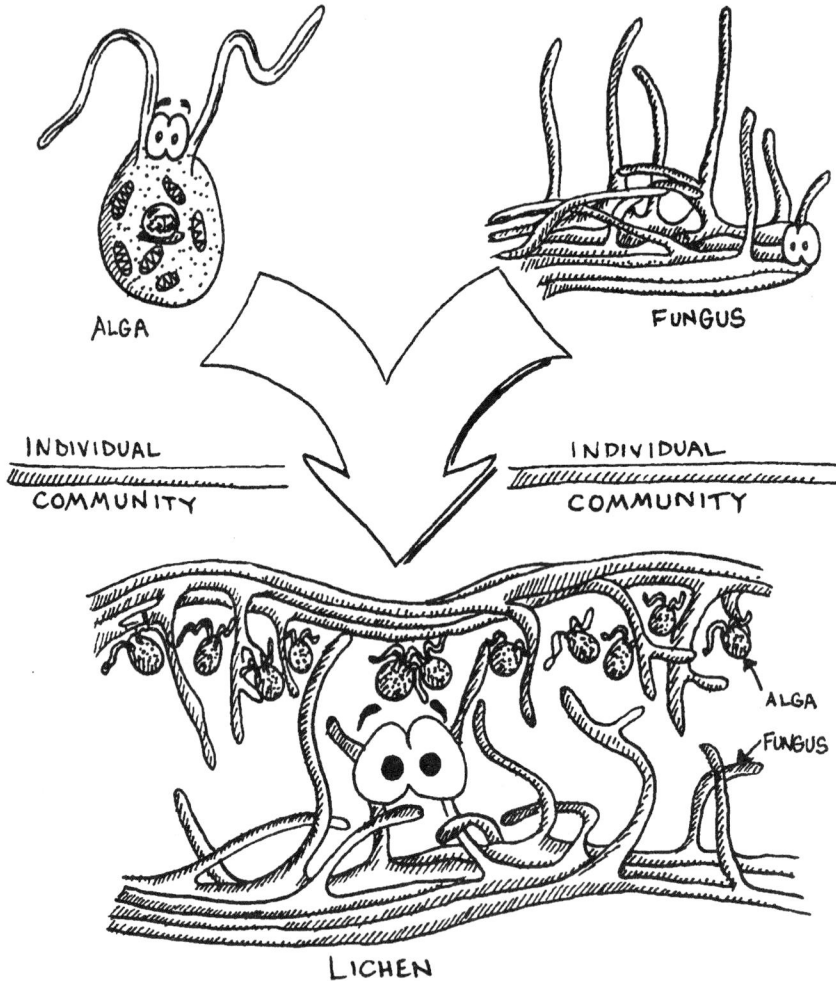

Figure 2.7 A lichen provides an example of Division of Labor at the community level. It is composed of an alga that associates with a fungus. When the two organisms associate, they become a new form of life that scientists recognize and classify as a distinct species.

？ Q & A with Jeff & Jen

JEFF: Interesting concept, the division of labor. I never realized how prevalent it is. Thinking back, I see that I began learning about it in freshman biology.

JEN: Right—way back during those first lectures on the eukaryotic cell. As a model of Ira's "Principle," it can't be beat. Volumes have been written about the cell. And furthermore, it's a great example of how the working parts of a system become mutually dependent. Some of the most tragic diseases known can be traced to a breakdown in the Division of Labor within cells.

JEFF: Really?

JEN: Yes. Take Tay-Sachs as an example. It's a genetic disease that's 100% lethal, and it occurs because the lysosomes fail to do their job.

JEFF: Lysosomes. Give me a second, here. Those are vesicles that contain an assortment of enzymes, right? They can digest virtually anything—the so-called "suicide bags" because if the lysosomes rupture, the cell can die.

JEN: Right. But they rarely break open like that. Their normal function is to digest old organelles and foreign matter that has entered a cell.

JEFF: People with Tay-Sachs don't have lysosomes?

JEN: They have them, but they're missing a crucial enzyme called hexosaminidase A. It breaks down a type of fat called the G_{M2} ganglioside.

JEFF: They lack a single enzyme—out of dozens—and that causes a disease?

JEN: A *lethal* disease. If hexosaminidase A is missing, death results—an indication of how precise the Division of Labor is.

JEFF: Why does it kill someone?

JEN: The G_{M2} ganglioside accumulates in brain cells because it isn't broken down. Nobody knows exactly why the fat is a problem, but the symptoms are consistent. A baby with Tay-Sachs is normal for the first six months, then starts to lose control of his arms and legs. It's usually blind by the first birthday, and almost always dies by age 4.

JEFF: Sad.

JEN: Agreed. But at least now it's possible for prospective parents to have themselves tested to see if they're carriers of the disease-causing allele of the gene.

JEFF: Some of this is coming back to me, now. Isn't this a disease that Jews have?

JEN: It's more common in them, but not exclusive to the group. Tay-Sachs appears in the Ashkenazic Jewish population at a frequency of about one in 3,600 births, and in the rest of the world's population at one in 400,000 births.

JEFF: Why in Jews?

JEN: The best guess right now is that it gives the carriers—heterozygotes that carry one copy of the Tay-Sachs gene and one copy of the normal gene—some measure of resistance to tuberculosis. And since Jews lived in dense ghettos for 2,000 years, where tuberculosis spreads quickly, the gene might have been valuable to their ancestors.

JEFF: This sounds a lot like the gene for sickle-cell anemia.[6]

JEN: There are a lot of similarities here. In both cases, a gene that creates problems had some benefit in the past. There are other diseases that aren't like that. For example, there is a rare disease in which a person's central vision suddenly deteriorates—wherever they focus their eyes, they only see a black spot, but they still have peripheral vision. It's called Leber's hereditary optic neuropathy. There isn't any apparent advantage to that, either now or in the past.

JEFF: Is it another breakdown in Division of Labor?

JEN: Yes, but this time in the mitochondria. There are 17 different mutations in the DNA of mitochondria that produce the effect. The mutations interfere with ATP production, and that somehow causes part of the optic nerve to quit working.

JEFF: So it's like Ira Bloomfield says: "One group quits, they all crash 'n' burn."

JEN: That's pretty close. The point is that the Division of Labor is so precise in a eukaryotic cell that a slight variation in any organelle can affect the cell as a whole. In the case of Leber's disease, this means that part of the visual field is lost.

JEFF: So, since I don't have any genetic diseases, I must be nearly perfect.

JEN: I know a lot of people that would laugh at that, Jeff.

---

6. See Chapter 1 for a discussion of this disease.

RELATED RESOURCES

Curse and blessing of the ghetto, J. Diamond, *Discover* 12 (March), 60–65, 1991.

Oxidative phosphorylation diseases and mitochondrial DNA mutations: diagnosis and treatment, J. Shoffner and D. Wallace, *Annual Review of Nutrition* 14, 535–638, 1994.

The wheels within wheels in the superkingdom Eucaryotae, A. Fisher, *Mosaic* 20 (Fall), 2–13, 1989.

Bees: Their Vision, Chemical Senses, and Language, K. von Frisch, Cornell University Press, Ithaca, New York, 1971.

CYCLING

Theme

Cyclic processes are seen throughout biology. In them, the components of a system return to their original states when a process is completed. Cycles are often linked to other cycles, so an interruption in one of them can have far-ranging effects.

Examples

- Enzyme action
- ATP synthesis and hydrolysis
- The Krebs cycle
- The cell cycle
- Heart contraction
- Double-circuit circulation
- Alternation of generations
- Population cycles
- The carbon cycle

As you read this, a holy man is pacing slowly around a Buddhist temple in Tibet, meditating upon a complex mix of theology and philosophy that he has studied since youth. Nearby, believers are spinning prayer wheels, clockwise, to call upon spirits that quicken their messages to the gods. What these people have in common with a disciple of biology is a fascination with the grand cycles of nature. They differ in that a biologist's knowledge arises from science, while the Tibetan's beliefs are based in faith.

The Tibetan brand of Buddhism recognizes many cycles of nature, and thus shares a theme with biology. The natural world—as explored by the scientific method—is replete with cycles. From the workings of molecules to the rise and fall of populations, cycles are as diverse as they are complex. And despite their differences, all cycles have a few things in common: something enters, something leaves, and the working parts of the cycle are reconstituted to begin the process again.

As an example, consider the **catalytic cycle of enzymes** (Figure 3.1). All cells are saturated with enzymes, for the simple purpose of enhancing reactions between other molecules, called the substrates. The catalytic cycle has four steps:

a. The substrates come into contact with the enzyme, and enter a region of the enzyme called the active site.

b. The shape of the active site changes slightly, fitting more tightly to the substrates. As a result, the substrate structure is altered. (In this case, a covalent bond is created. Enzymes can also break bonds.)

c. In its new state, the substrate escapes from the active site.

d. The enzyme reverts to its original shape, and is ready to catalyze the reaction again.

The fourth step represents the essence of a cycle—the enzyme returns to its initial state. What would happen if the fourth step was not completed? A cell would have to make a new enzyme each time the reaction was needed. Because it is reused thousands of times, an enzyme saves the cell an enormous cost.

Few examples of cycles are so streamlined as this one, as you will learn. But in all natural cycles, the four steps can be discerned. In generalized form they go like this:

Step 1: An object enters the cycle.

Step 2: The components of the cycle interact with the object, changing it in some way.

Figure 3.1 The catalytic cycle of a typical enzyme. (A) An enzyme contains a region called the active site, which is receptive to specific molecules called the substrates. (B) When the substrates enter the active site, the structure of the enzyme alters slightly. This change, called the induced fit, brings the substrates into the proper orientation for a reaction. (C) A reaction takes place. (D) The substrate molecules, now altered in some way, leave the active site. Since the enzyme has returned to its original shape, it is poised to go through the cycle again.

Step 3: The object leaves the cycle.

Step 4: The components return to their original states, and the cycle begins again.

Why are cycles so widespread? Why are they so important in biology? These are two of the questions that will be addressed in this chapter. And we will find that the answer depends on the level of The Living Staircase where the cycle is operating. From the level of molecules to individuals, cycles function to increase metabolic and reproductive efficiency, leading to greater survival and reproduction. From the population level on, however, it is hard to envision a function; cycles seem to occur for no particular reason except that they are the way things work.

To a Buddhist, the grandest cycle of all is that of life and death—the cycle of reincarnation. Its function is to improve the spirit. In their view, one's life is analogous to a substrate within the active site of an enzyme. The spirit enters a newborn (the active site), is modified by a lifetime of unique experiences, then leaves (death). After going through a long sequence of reincarnations, spiritual perfection is achieved. Tibetans believe that at any point in time, there is only one person on the planet that has achieved the perfect state—the Dalai Lama, leader of the faith.

The ATP Cycle

It is intriguing that, to contemplate Tibetan beliefs, a molecular cycle has to take place in our brains. The same cycle is required for a Lamaist to spin a prayer wheel—he or she moves the fingers in a precise, yet subtle sequence of contractions—contractions that are fueled by molecular activity inside the muscles themselves. I am speaking of the hydrolysis and reconstitution of ATP (adenosine triphosphate), which is carried out with the aid of cellular organelles known as mitochondria.

Biology students around the world learn that ATP is the main source of power for the chemical reactions of a cell, but they seldom grasp the cyclic mechanism by which ATP is reused thousands of times. A single molecule breaks up to release energy, then comes back together. To begin, let's examine the part of the cycle in which energy is released.

Every ATP molecule contains a high-energy phosphate group that can be transferred to energize another molecule. At the time of transfer, the receiving molecule is poised to undergo a chemical reaction, but does not have sufficient energy. Receipt of the phosphate group increases the potential energy of the reactant, and when the phosphate bond breaks, the chemical reaction goes to completion (Figure 3.2). This is the first phase of the ATP cycle—the energy-bearing ATP molecule (adenosine **tri**phosphate) is converted from a single unit into two pieces: ADP (adenosine **di**phosphate) and an **inorganic** phosphate group that has lost its high-energy bond.

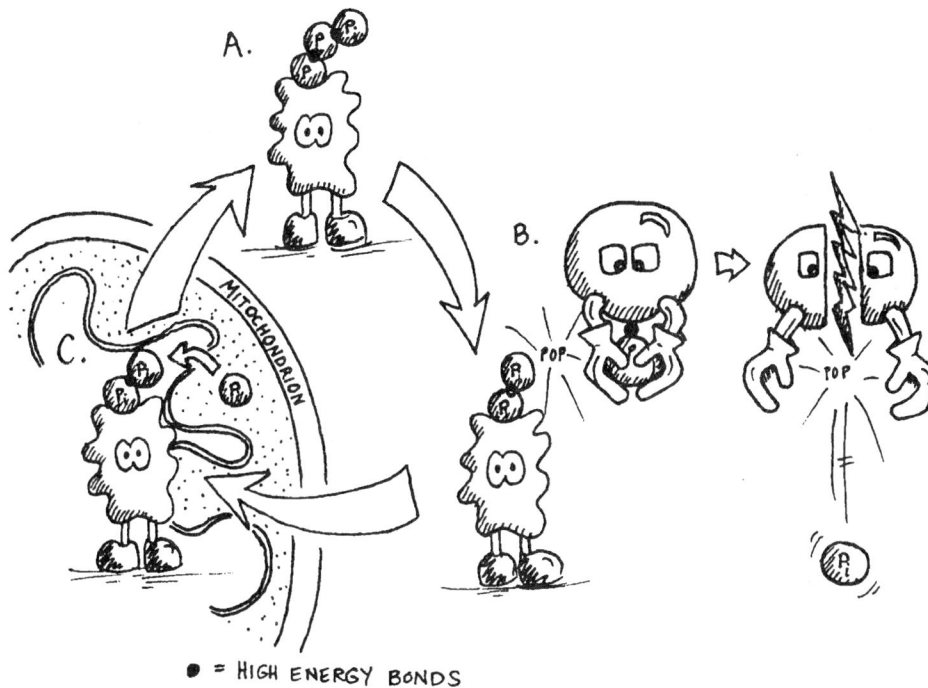

● = HIGH ENERGY BONDS

Figure 3.2 The ATP cycle occurs constantly in the cytoplasm of all cells. (A) The ATP molecule contains three phosphate groups, two of which have high-energy bonds. (B) In most cases, only one of the phosphate groups is transferred to a reactant. After being transferred, the phosphate bond breaks, releasing energy to force a reaction (in this case, splitting of the substrate). The phosphate group, now independent, floats away. (C) ADP is rejoined with a phosphate group at the inner membrane of a mitochondrion. ADP can be linked to any phosphate that is available.

Life thrives on this process—it is the main source of energy for the reactions that occur in cells. In fact, you had to hydrolyze (break) more than a trillion ATP molecules to move your eyes through this paragraph.

In the second part of the ATP cycle, ADP is reunited with an inorganic phosphate. If this did not occur, a cell would quickly run out of energy. For example, a muscle cell typically contains about 500 million molecules of ATP, and runs through the whole supply within a minute. Without a recycling mechanism, only 60 seconds of activity could be sustained. So how are the ADP and inorganic phosphates reunited? There are several ways, but the most common one involves mitochondria.

A mitochondrion is like a recycling plant that receives shipments of empty bottles and plastic caps at the perimeter, and with machinery of amazing complexity on the inside, restores the caps (P) to the bottles (ADP), while filling them with energy (the high-energy phosphate bond). The drinks are shipped away to be drained, and afterward, the bottles and caps are returned to the plant.

The inner membrane of a mitochondrion houses an enzyme, ATP synthase,[1] whose active site accepts ADP and inorganic phosphate. With these substrates in place, it is poised to make ATP. There is an energy cost, however. (Since the breakdown of ATP releases energy, the reunion of ADP and P must have a cost.) To meet the cost, the mitochondrion runs its own cycle, **the Krebs cycle** (Figure 3.3).

Cycles within Cycles

Remember that, for any cycle, something enters, is modified, and then leaves. The Krebs cycle (also known as the citric acid cycle and the tricarboxylic acid cycle) is a series of eight chemical reactions, each catalyzed by a specific enzyme. An **acetyl group** enters the cycle—it is a two-carbon molecule containing oxygen and hydrogen. As the Krebs cycle progresses through its reactions, the acetyl group is completely dismantled, and its components exit separately. Carbon and oxygen leave as CO_2, and the hydrogens exit in association with carrier molecules (NAD^+ and FAD) that become NADH and $FADH_2$.

The underlying purpose of the Krebs cycle is to supply energy to ATP synthase. This is accomplished when NADH and $FADH_2$ deliver the hydrogens to the inner membrane of the mitochondrion, where the electron transport system (ETS[2]) is operating. ATP synthase is positioned close to the ETS, and receives energy from it.

Let's stand back and take a panoramic view of the overall process. The mitochondrion's role in the ATP cycle is to link ADP and inorganic phosphate. It contains an enzyme, ATP synthase, that can do this directly. However, there is a substantial energy cost. To supply the energy, the mitochondrion runs the Krebs cycle, which produces hydrogens that are delivered to the electron transport system. The ETS provides energy for ATP synthase.

A mitochondrion illustrates an important point—it often takes a cycle (the Krebs cycle) to maintain a cycle (the ATP cycle). The two are linked like the gears inside a mechanically driven clock—the cogs of the Krebs cycle interlock with the ATP wheel, ensuring that it continues to turn. And to carry things further, the Krebs cycle is itself driven by other wheels—those of the eight enzymes that make up the Krebs cycle itself. Each enzyme, considered alone, is continuously going through its own catalytic

---

1. The thermodynamic properties of this enzyme are discussed in Chapter 8.
2. The electron transport system is described in Chapter 8.

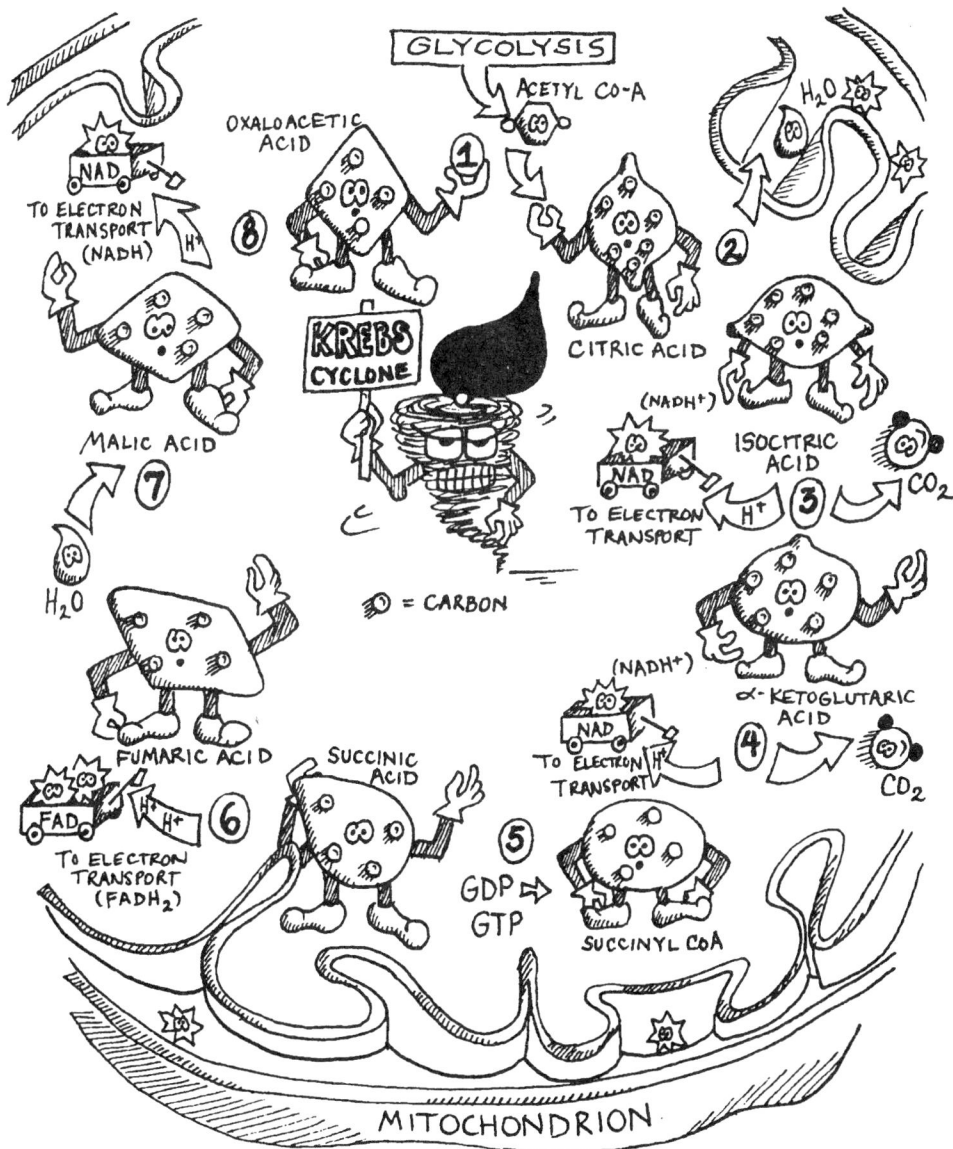

Figure 3.3 The eight-step Krebs cycle occurs in the interior of a mitochondrion, and supplies hydrogens to the electron transport system. In the first step of the cycle, acetyl CoA donates its two-carbon acetyl group, converting a four-carbon acid into a six-carbon acid. Hydrogens are released at steps 3, 4, 6, and 8, and carrier molecules (NADH and FADH2) move them to the electron transport system. At steps 3 and 4, two carbons are released as CO_2, leaving four carbons within the cycle. These carbons are modified to become oxaloacetic acid again. Note that the reactant of step 1 is the same as the product of step 8—thus the cycle is complete.

cycle (Figure 3.1). We have only begun to explore The Living Staircase, and already we find cycles within cycles. It is enough to inspire the Dalai Lama himself.

The Cell Cycle

Molecular cycles commonly interact, as the above example shows. But they also influence higher-level cycles. The life of a cell, which is a cycle in itself, is a good example (Figure 3.4). At the beginning of its life, a cell goes through a growth phase, called G_1. Then, it pauses during the S phase, when the DNA molecules are copied. After completing a second growth phase, G_2, the cell divides to finish its life cycle. At the end, there are two new cells, both of which are identical to the original.

How does a cell "know" when it has completed one part of the life cycle and is ready for the next? Apparently, it uses molecular cues.[3] One such cue involves an enzyme known as maturation-promoting factor (MPF), which stimulates the cell to leave the G_2 phase and begin the cell division phase (Figure 3.4). In a typical cell, MPF increases slowly during the second growth phase until it reaches a "threshold concentration." At this point, it stimulates the cell to divide by activating a host of other proteins. As cell division takes place, the concentration of MPF falls rapidly. Thus the daughter cells start out with the same concentration of MPF that the parental cell had at the beginning of its life. A molecular cycle (the rise and fall of MPF) has had a major influence on a cellular-level cycle.

The cell cycle is fundamental to life. All humans start out as a single cell, and by the continuous operation of the cell cycle arrive at full size, composed of trillions of cells. At adulthood, most of our cells exit the cycle and go into a stable state called G_0. But even then, the cell cycle plays a crucial role in sustaining life—it is the mechanism that replaces cells that have been lost. As an example, consider the cells that we lose every day, just by making contact with the world—the skin.

The skin is analogous to the shingles that protect a roof from the weather. It is composed of dead cells stacked on top of each other, dozens of layers deep (Figure 3.5). Over the course of a day, the outer surface erodes, but the skin does not become thinner because new layers are continuously produced from below, in a region called the stratum germanitivum. Here, the cell cycle runs at a rapid pace. Each cell in the stratum germanitivum grows, divides, and sends one of the daughter cells to the surface. The rising cell is slowly cut off from its blood supply, so it dies. But what about the other cell, the one that did not leave the stratum germanitivum? It goes through the cycle again: nutrients enter, the cell divides to produce two new cells, and one of the cells leaves the cycle for the surface. The remaining cell begins anew.

---

3. Molecular communication is the subject of Chapter 9.

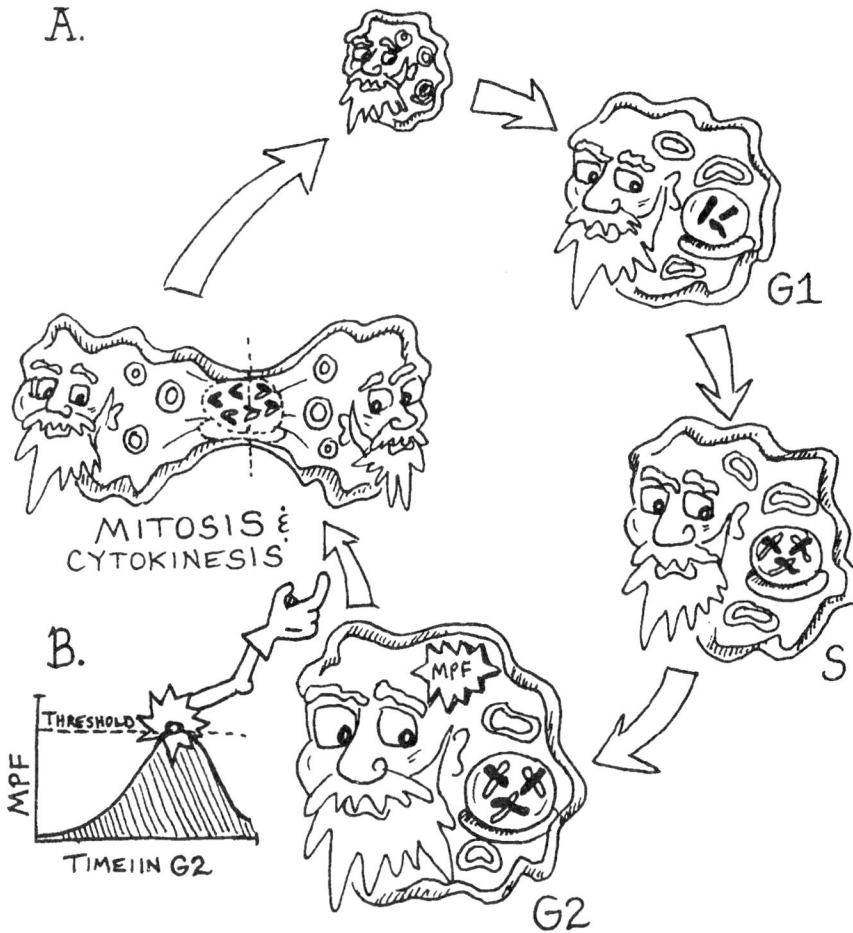

Figure 3.4 The cell cycle operates continuously in all but a few organisms. (A) In the gap 1 (G_1) phase, the cell grows and adds to its complex of internal membranes. In the synthesis (S) phase, each chromosome replicates, so the amount of DNA in the nucleus doubles. In the gap 2 (G_2) phase, the cell continues to grow. When it reaches a size that is about twice as large as its initial size, it enters mitosis. (B) Production of MPF begins in the G_2 phase. When MPF reaches a threshold level, it stimulates the cell to begin mitosis.

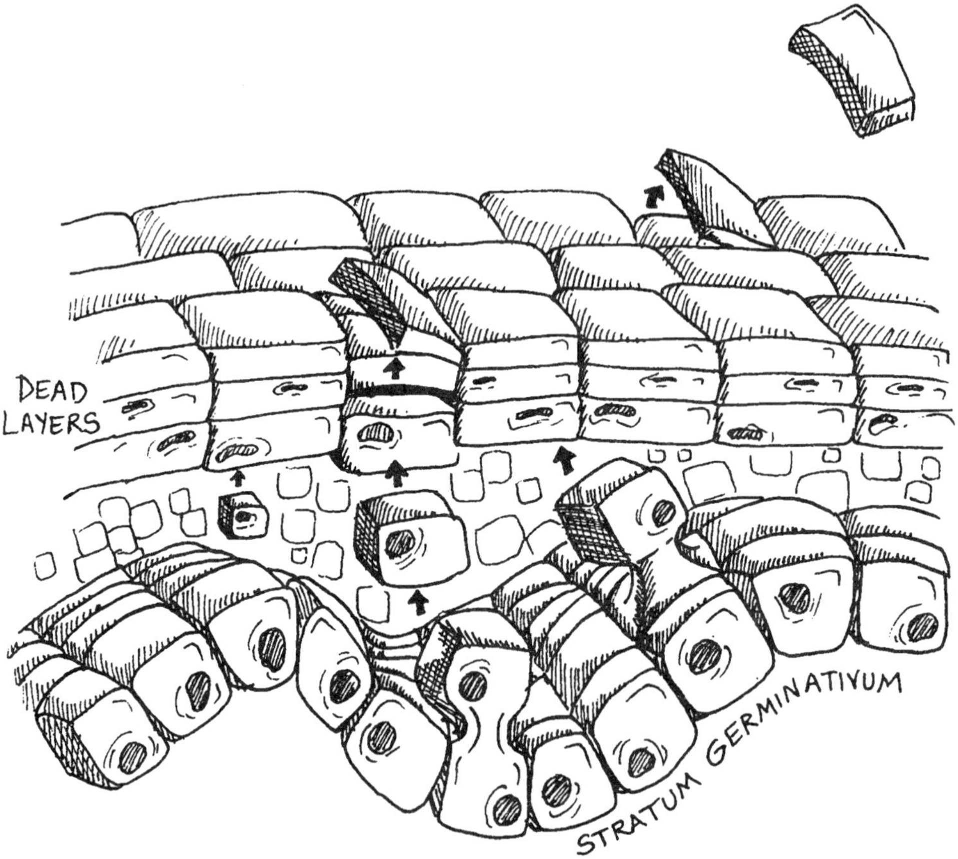

Figure 3.5   The upper layers of the skin are composed of dead cells that arise from a deeper layer, the stratum germanitivum. In the living layer, the cell cycle operates continuously, each cell dividing into two daughter cells. One of the daughter cells is pushed upward, to become the dead protective layer, while the other daughter cell remains in the stratum germanitivum to continue dividing. By retaining one cell to continue the cell cycle, the stratum germanitivum can replenish the skin indefinitely.

This is a classic example of Division of Labor[4]—one cell is devoted to continuing the cycle while the other is added to the protective layer. And it is essential for life to continue. What would happen if both daughter cells left the stratum germanitivum for the surface? The skin would wear away, like a roof does. By retaining one cell to begin the cycle again, the stratum germanitivum can resupply the skin indefinitely.

---

4. See Chapter 2 for more examples.

Theoretically, the stratum germanitivum could keep doing its job forever. However, it eventually ceases to exist—all animals die, and take perfectly good cells along with them. We put a good deal of energy and effort into a terminal existence. To a Lamaist, however, the energy and effort of staying alive are investments—the spirit benefits from them, and continues its quest. After being liberated at death, it goes through a waiting period (samsara), then enters another newborn. It will learn a new lesson in this life, and in the next, and the next.

Wouldn't it be better never to die in the first place? This is a fantasy for humans, but a reality for the "lowly" prokaryotic cells of kingdom Monera. Consider a single-celled bacterium. It grows, then divides into two identical cells—it is reborn in duplicate, one might say, and begins two new cycles of life. (There is nothing like this in animals; we do not "become" new copies of ourselves. Rather, we produce offspring that are a combination of ourselves and another person.) In each generation, some bacterial cells die, but others survive and become the next generation. Thus, every bacterial cell that lives in your gut today has a continuous history that stretches back at least 200,000 years (as long as humans). The immortality that the Tibetan monk meditates upon is forbidden to the human body, yet is routine for the simplest form of life on earth!

The Heart Cycle

Thus far in our exploration of The Living Staircase, we have reviewed cycles at the level of molecules (ATP, MPF, the catalytic cycle of enzymes, and the Krebs cycle), cells (the cell cycle), and tissues (the cycle of the stratum germanitivum). As we move along The Living Staircase, cycles are everywhere. Let's move on to organs. One cycle at this level is a source of periodic embarrassment. After we eat a meal, the stomach squeezes down on its contents roughly three times a minute, and each contraction is a cycle of work and relaxation in itself. When we skip a meal, the contractions squeeze down on air, creating the audible growls that amuse those around us. But among organs, perhaps the most intriguing cycle is that of the heart. It is an inexhaustible worker, a stranger to rest. From before we are born until the day we assume room temperature, the heart cycles continuously, precisely, faithfully.

The heart of a human is composed of four chambers—two atria positioned on top of two ventricles—that coordinate their contractions to keep the blood moving in the right direction. The atria contract at the same time, and after a short delay, the ventricles contract in unison. This is why as embryos we hear our mother's heartbeat as two quick thumps, instead of four thumps that would result if each chamber contracted on its own.

At the beginning of the heart cycle, blood enters the atria via large veins (Figure 3.6). After filling, the atria contract to force the blood into the ventricles below. When the ventricles contract, the blood is pumped into arteries that lead to every region of the body. In a person at rest, one cycle of contractions is completed every eight-tenths of a second, producing about 70 heartbeats per minute.

A Tissue Cycle Drives the Heart Cycle

The heart cycle has a clear purpose. If the blood is to move in the right direction, the atria **must** contract before the ventricles. Unsynchronized contractions result in heart attacks. Precise synchrony, then, is the goal of the cycle.

How are the contractions synchronized? In the right atrium there is a cluster of muscle cells called the sinoatrial node (or pacemaker) that are myogenic. In other words, they initiate their own contractions. The heart cycle of a resting human is 0.8 seconds long because the cells of the sinoatrial node contract at that frequency. With each contraction, these cells generate a pulse of electrical activity that radiates outward very quickly, stimulating other muscle cells to contract. The impulse moves so fast that the cells making up the atria contract at virtually the same time.

Fast transmission of information is usually beneficial, but in the heart it presents a problem. If the impulses from the sinoatrial node arrive at the ventricles too quickly (i.e., before the ventricles fill with blood), the ventricles will transport only a portion of the blood that they are capable of pumping. To handle this, there is a second cluster of cells at the lower end of the atria called the atrioventricular (AV) node. The cells of the AV node receive the electrical impulse from the sinoatrial node, but delay its continuation into the ventricles by one-tenth of a second. This allows the ventricles to fill completely before they contract. Thus the cycle of the heart—a whole organ—is timed to perfection by two specialized tissues (the sinoatrial and AV nodes) that experience their own cycles of contraction and relaxation. Like the mitochondrion, the heart depends on cycles within cycles.

The Heart Cycle Drives Yet Another Cycle

The heart cycle creates the force to run yet another cycle, one that is a step away along The Living Staircase, at the level of systems—the cycle of circulation. Let's begin with the right half of the heart. Oxygen-depleted blood is forced into the right ventricle when the atrium above it contracts. From this ventricle, the blood is forced into the pulmonary artery, and the cycle of circulation begins (Figure 3.6).

Propelled by the contraction of the right ventricle, blood in the pulmonary artery rushes toward the lungs. There, it travels through capillaries and picks up the oxygen that is essential to every cell in the body. Oxygenated blood then drains into the pulmonary vein, which leads back to the heart. But instead of returning to the right

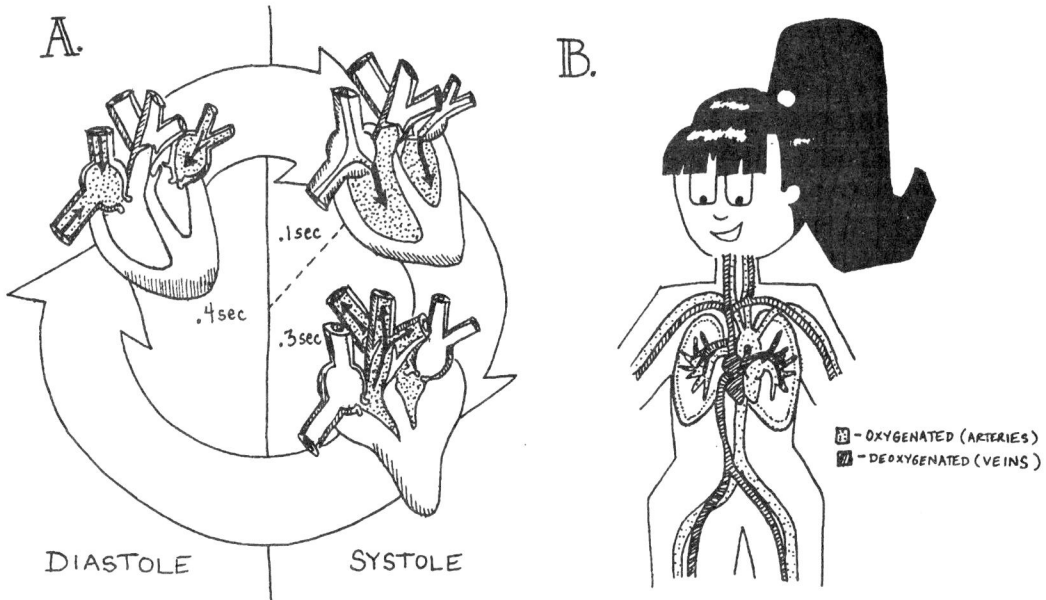

Figure 3.6   The heart cycle provides the force to drive the circulation cycle. (A) The heart cycle in a resting person. In the first 0.1 seconds, the atria contract to force blood into the ventricles. Next, the ventricles take 0.3 seconds to contract, sending blood into the major arteries. In the remaining 0.4 seconds, the atria refill. (B) The cycle of circulation. Blood leaving the right ventricle travels to the lungs to pick up oxygen. After returning to the heart, this blood is pumped by the left ventricle to the tissues, where oxygen is delivered. Note that one cycle at the organ level (the heart cycle) drives the next cycle at the system level (the circulation cycle).

side, the blood coming from the lungs enters the atrium on the left side. At this point, the cycle of circulation is half done (Figure 3.6).

From the left atrium, the oxygen-rich blood is forced into the left ventricle, the most powerful of the four chambers of the heart. When the left ventricle contracts, blood rushes into the aorta, which branches to carry blood to each organ. Inside an organ, blood flows through capillary beds, and the ultimate goal of circulation is accomplished—tissues near the capillary bed receive oxygen and nutrients, and metabolic wastes are carried away. From the capillary beds, blood drains into veins which lead back to the heart, completing the cycle.

When one looks at the overall scheme of circulation in Figure 3.6, it gives the appearance of being a two-cycle process—in the first phase, blood travels to the lungs to pick up oxygen, and in the second, the oxygen is delivered to needy tissues. And in fact, circulatory systems of this nature have two circuits; but the term "circuit" has a different meaning than "cycle." By the definition of a cycle, something (oxygen) must enter (at the lungs), then leave (at the tissues); thus the two **circuits** comprise a single **cycle**.

The details of circulation point out a key feature of biology, and the main point of this book. It is this: In studying about life, we separate organisms into pieces to make the job easier. But in doing so, we have to bear in mind that the subdivisions are artificial—this approach is like separating the plumbing, air conditioning, and framework of a building in an attempt to understand how it functions as a whole. If we neglect to put the pieces together—at least mentally—we will fail to get a complete picture. The circulatory system shows us three cycles operating simultaneously to achieve the relatively simple goal of getting oxygen to the tissues: (1) the cyclic production of electrical impulses by nodes in the heart; (2) the heart cycle at the organ level; and (3) the cycle of blood circulation at the systems level. In the same way that a Lamaist (1) spins a prayer wheel while (2) pacing around a temple on (3) a revolving planet, cycles of nature operate simultaneously at different levels.

The Cycle of Reproduction

Even the life of a whole animal can be described as a cycle.[5] Consider the reproductive life of a woman. About every 28 days, she produces a single cell that has the potential to become another human. If the cell is not fertilized, she resumes the cycle to produce another one four weeks later. If the cell is fertilized, she enters

---

5. Chapter 9 discusses the voluminous amount of communication required for the cycle of reproduction.

a different cycle—the embryo grows in the uterus, is born, then nurses for some period of time before she returns to the original 28-day cycle.

The cycle of reproduction in animals is fairly simple, but the cycle of reproduction in plants is complicated by having two distinct phases, and requiring two generations to complete. Rather than being experienced by an individual organism, it is a species-level cycle that requires two generations. Called the **alternation of generations** (Figure 3.7), it is a source of confusion to students of biology everywhere.

Ferns provide a good example. Those that we notice in the wild are diploid organisms (i.e., the cells contain two copies of each chromosome), and are referred to as sporophytes. (The name reflects the type of cell that the plant will use for reproduction—spores.) When the right season comes around, microscopic chambers develop on the undersides of the leaves. Each chamber contains a group of cells that will develop into spores, and each spore will have one copy of each chromosome (i.e., the spores are haploid). Thus, a diploid plant produces offspring that are haploid, and one generation of reproduction is not a complete cycle. (Remember that in any cycle, the end point is equivalent to the beginning.) The haploid spore must produce diploid offspring before the cycle is finished.

And that is exactly what happens. On a warm and dry day, ferns in the sporophyte generation release their offspring to the wind, and the spores eventually settle. The majority of them die because they land in an unsuitable habitat, but a few end up in a moist, shady place where the spores can grow. They develop into a plant that few people would recognize as a fern. It is a flat, heart-shaped structure, about the size of a dime, which resembles a small leaf. Called the gametophyte generation, it uses sperm and egg cells for reproduction. After a rain, or when enough dew collects on the ground to immerse the gametophyte in a film of water, the sperm cells swim over and fertilize the eggs of another gametophyte. (Within a gametophyte, the sperm and eggs mature at different times so that self-fertilization cannot occur.) The resulting zygote is diploid because it has received two sets of chromosomes—one from the sperm cell and one from the egg—and it develops into a new sporophyte. After two generations, the cycle of the fern is complete.

While the alternation of generations is a species-level cycle, it is important to understand that both the sporophyte and gametophyte generations have their own, individual cycles of reproduction. A sporophyte plant lives for years, and goes through a reproductive phase annually, always producing spores. The gametophyte generation has a shorter life span, but even so it has at least one reproductive cycle. It is only in combination that these two generations complete the species-level, diploid-to-haploid-to-diploid cycle.

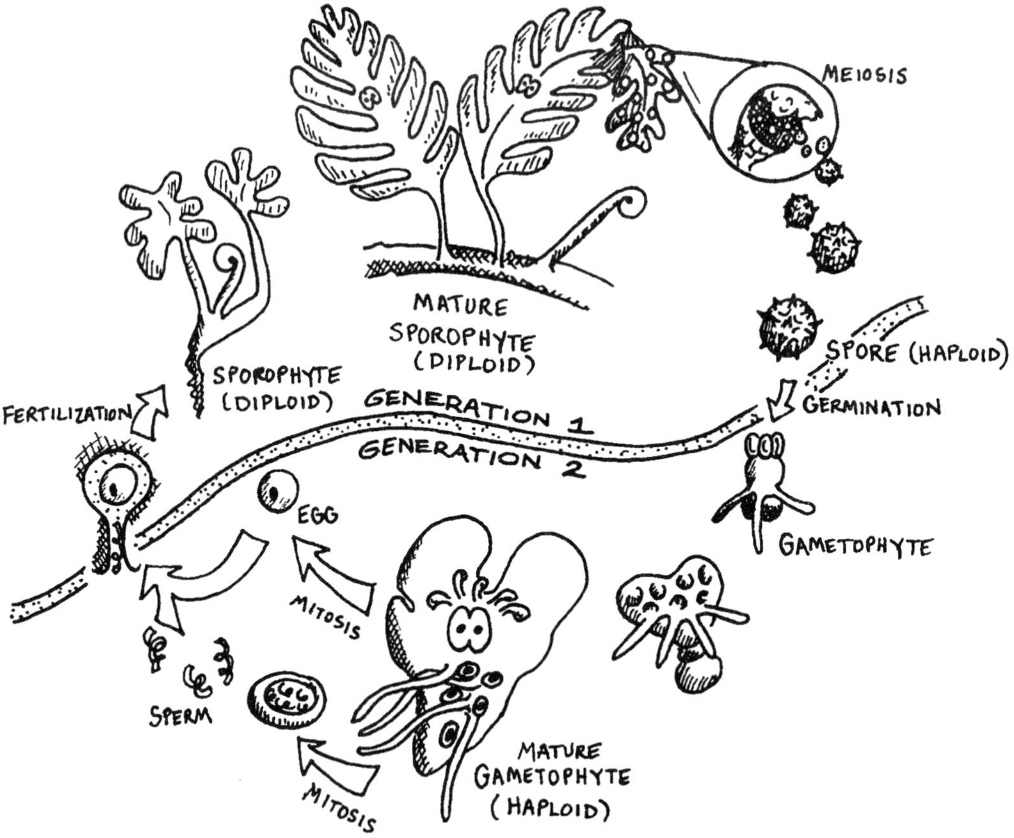

Figure 3.7 The alternation of generations in a fern takes two generations to complete. The sporophyte generation is diploid, and uses meiosis to produce haploid spores for reproduction. These spores germinate to become the next generation, the gametophyte. Gametophytes produce sperm and egg cells for reproduction. When the gametes unite, the two-generation cycle is complete.

The Value of Cycles

Natural selection is very good at explaining the advantage of cycles from the level of molecules to the level of individuals. An efficiently operating Krebs cycle enhances the growth rate of an organism's cells; the cell cycle in the stratum germanitivum replenishes the skin, and thus provides a barrier to pathogens; a carefully regulated heart cycle keeps the tissues alive and functioning. All of these cycles, taken together, prolong the life of the organism and enhance the cycle of reproduction. Thus the cycles that characterize the lower half of The Living Staircase (from molecules to individuals) all work toward the same end—enhancing the ability of the organism as a whole to reproduce.

When we go beyond the level of the individual, however, we encounter cycles that are a greater challenge to explain. At the species level, the alternation of generations in plants is a good example. What is the advantage of a two-generation reproductive cycle? Cycles also occur at the population, community, and ecosystem levels. How are these adaptive? How do they enhance reproduction? In all likelihood, they don't. And thus they require a different type of explanation. Before attempting such an explanation, let's look in detail at another example—the population cycle of lemmings.

A Challenge to Explain: The Lemming Cycle

Lemmings are small brown rodents about the size of hamsters (Figure 3.8). Their habitat is the northern tundra, where vegetation is scarce and the climate is severe for most of the year. Throughout this region, lemmings experience large-scale fluctuations in population density. This is not surprising, given the harsh climate. But there is a perplexing feature of the cycle—it consistently takes four years to complete (i.e., four years pass between successive population peaks). When a lemming population has a typical peak, it attracts the attention of predators and naturalists. But occasionally, a population grows to such an extraordinary density that it attracts journalists (who by some accounts are close relatives to rodents). At such times, lemmings have been reported to madly rush over one another, sometimes diving over cliffs in massive numbers to end their miserable lives. These are exaggerations, but one thing is true—nobody overlooks one of the truly impressive lemming highs.

To date, there is no single, well-supported explanation for the four-year duration of the lemming cycle. But there is no shortage of ideas. One hypothesis invokes a second cycle—an underlying nutrient cycle in the food chain that lemmings depend upon (Figure 3.8). Let's begin with a small population that is growing rapidly. In this stage, the lemmings are healthy and vigorous because the vegetation they feed upon is nutritious. Population growth continues for a year or two.[6] But as the population expands, it reduces the supply of available food. Eventually, snow arrives to find the lemmings far more abundant than the winter food supply, and the population moves into crisis. Some lemmings starve; many more take their chances by migrating; and others stick out the tough conditions in place.

If the lemmings' food supply could recover the following spring, the rodent population would rebound and show an annual cycle. Such a cycle would not be unusual; most animals have a "summer high, winter low" pattern of abundance. The

6. The way that male lemmings choose whether to mate with a single female or a number of females may be controlled by a single hormone. See Chapter 9 for the details.

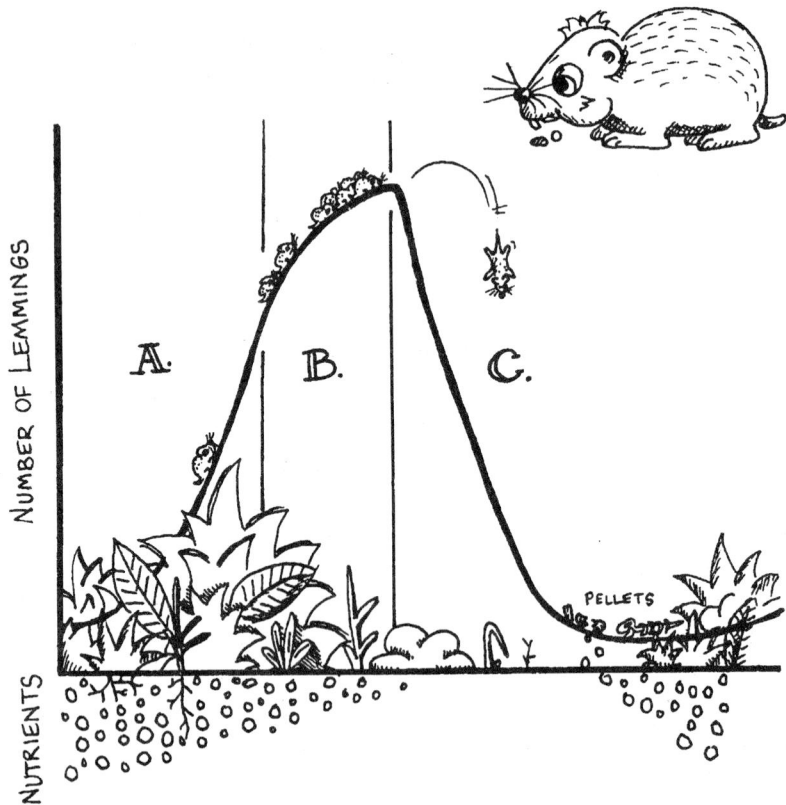

Figure 3.8   The lemming cycle of arctic regions may be driven by a cycle of nutrient availability in the soil. (A) When the soil is rich in nutrients and vegetation is abundant, the lemming population grows quickly. (B) As lemmings become more numerous, they deplete the vegetation and transfer the nutrients from the soil into their fecal pellets. The vegetation suffers because the soil is poorer. (C) At peak densities, vegetation is scarce and the soil is nutrient-deficient. The lemming population crashes. It remains low until the lemming pellets decompose, returning the nutrients to the soil. Note that if lemmings were not present, the cycle of soil nutrients would not occur.

lemming cycle is unique because there is a period of four years between the growth phases. And for lemmings, this is where the hypothesized nutrient cycle comes in.

When the dense lemming population is consuming the food supply, nutrients are moving through the ecosystem in a predictable sequence. First, they are absorbed from the soil and incorporated into plant tissues. When eaten by a lemming, some of the nutrients become animal tissue, and some simply pass through a lemming's body to be stored in a fecal pellet on the surface of the soil. The important point is that the nutrients are transferred from the soil, where they were available to roots, to the ground surface where they are not. In the following summer, the soil is poor, vegetation is scarce, and the rodent population crashes.

How long will the population be kept low? According to this hypothesis, lemmings will remain scarce until the pellets decay, returning their nutrients to the soil. In the arctic climate, summers are brief and temperatures are cool; the breakdown of pellets apparently takes a couple of years.

Now let's step back and ponder the question we began with: What is the purpose of such a cycle? Certainly it isn't in the interests of the little buggers—struggling to survive for three years and starving in the fourth isn't by any stretch a good adaptation.

So why do lemmings cycle? If the hypothesis is on target, the cycle is driven by *extrinsic* factors, i.e., by things that are not built into the genes of the animals. The lemming cycle may exist for the simple reason that a growing population will deplete nutrients in the soil, and it takes a few years for the nutrients to return. They do not cycle because it is adaptive—they cycle because, like most organisms, they do not forecast the effects that their activities will have.

The Global Cycle of Carbon

As a supposedly intelligent species, we may hope to learn from the shortsighted-ness of lemmings. Many of the environmental problems we face today seem to have arisen from a human influence on large-scale cycles. Ten thousand years ago, the human population was so small that our ancestors could not have affected the biosphere even if they had tried. But because of an extraordinary rate of population growth, we can now affect the natural environment just by going about our normal activities. To have a good understanding of why this occurs, one has to comprehend the grandest cycles of nature—the biogeochemical cycles that take place at the top of The Living Staircase.

We will follow the carbon cycle as an example. Carbon is the most common element in living things, making up about 48% of all cells, and yet it is surprisingly scarce on earth. The source of carbon for all life is the atmosphere, where it exists as carbon dioxide. When you inhale, the air entering your lungs is 78% N_2, 21% O_2, and 0.03% CO_2. (Other rare gases that are not important to living systems make up the remaining 0.97%.) Furthermore, this scant amount of CO_2 is inaccessible to animals; only the photosynthetic organisms—plants, some protistans, and some bacteria—are capable of taking CO_2 from the atmosphere and using it to construct biological molecules.

Figure 3.9 A carbon molecule cycles through an ecosysytem over a period of about a decade. (A) The source of carbon is CO_2 in the atmosphere. CO_2 is picked up by photosynthetic organisms at the beginning of the Calvin cycle; rubisco is the enzyme that links CO_2 to an existing organic molecule. (B) After becoming part of a plant, a carbon molecule may travel through any part of the food web as one organism eats another. (C) Carbon returns to the atmosphere via the Krebs cycle. Since all aerobic organisms operate the Krebs cycle, any species can return carbon to the atmosphere.

How do they tap into the supply of CO_2? All photosynthetic organisms have the enzyme known as rubisco (ribulose biphosphate carboxylase). Rubisco catalyzes a reaction known as carbon dioxide fixation, and in doing so, it brings carbon within the reach of all living things (Figure 3.9). At the moment that the reaction occurs, a molecule of carbon dioxide—one that may have been floating around idly for thousands of years—forsakes the security and boredom of the atmosphere to become a component of a living cell. This is the first step of **the carbon cycle**.

Rubisco adds the CO_2 molecule to a five-carbon acid, producing a six-carbon product. Through a series of reactions known as the Calvin cycle, the six-carbon molecule is modified, and glucose is ultimately produced.

Upon becoming part of a glucose molecule, a carbon atom can go along several paths. The glucose might be broken down by a mitochondrion; in that case, the carbon atom is released as CO_2 to the atmosphere again. The glucose might be linked to other glucose molecules to make cellulose; or the glucose might be modified to become some other kind of molecule in the cell of, say, a leaf. If this leaf is eaten by an insect, the carbon becomes part of the insect's body; and if the insect is eaten by a bat, the carbon atom's adventures continue. By being passed from one species to another along the food chain, a single carbon atom can persist in an ecosystem for decades, perhaps centuries. This is a concept that a Buddhist could appreciate: The carbon atoms that make up 48% of a human come from any number of places, and would have most remarkable histories, if only we could trace them. But they all have one thing in common—at some time in the past, each one was picked up from the atmosphere by rubisco and used to make glucose.

Eventually, every carbon atom in a cell will escape to the atmosphere as CO_2. While there was only a single entrance point (rubisco), there are several exits. The main one, however, is a cycle that was covered earlier in this chapter—the Krebs cycle (Figure 3.2). The two carbons that enter the Krebs cycle as an acetyl group (which is a derivative of glucose) are channeled through a series of enzymes; and ultimately, they escape to the atmosphere as CO_2 where, by Tibetan accounts, the spirits of humanity are waiting to move back into the cycle of life.

It is an awesome thing to ponder. Each time I exhale, I am losing CO_2 molecules that have experienced untold adventures in the biosphere. I may have just lost an atom of carbon that entered the realm of life 100 million years ago, when rubisco snatched it from the atmosphere and put it into a tree fern. It may have listed in its life story a stint as part of a plesiosaur, or as part of the hand that Napoleon tucked inside his uniform. It almost makes me hold my breath.

Notice that the carbon cycle, grandiose as it is, is driven by two subcellular cycles: the Calvin cycle bringing carbon into living tissues, and the Krebs cycle allowing carbon to leave. While in the atmosphere, CO_2 is said to be in its "abiotic reservoir," meaning that it is available to life, but is not directly connected to any living thing. However, in this abiotic state it is still important to the biosphere because it performs a crucial function for all life on the planet—that of keeping the earth warm.

Compared to the other eight planets in our solar system, the earth is unusually warm—warmer than it should be, given its distance from the sun. The average temperature of earth is appropriate to keep most of the planet's water in a liquid state, and this is believed to be a basic requirement for life to flourish. Carbon dioxide makes up only 0.03% of the atmosphere, but even in this small quantity it serves as an important insulator. Carbon dioxide has the right molecular structure to absorb the sun's rays, and thus retain much of their heat.

For any cycle to be "balanced," the rate at which materials enter has to be equal to the rate at which they leave. The cycles that occur at the lower levels of The Living Staircase are closely regulated so that this is almost always the case.[7] But at the population level and beyond, regulation is not so precise. For the carbon cycle, the inputs and outputs may balance when summed over eons, but over a shorter period they may not. At the ecosystem level, the proverbial "balance of nature" is more a myth than a reality.

During the reign of dinosaurs, for example, the entryway for carbon was much busier than the exit. This was a time when the climate on earth was warmer, and plants flourished—the Calvin cycle removed CO_2 from the atmosphere faster than the Krebs cycle could replace it. Many plants and animals did not decompose after dying; instead, they were buried under sediment and became coal and oil.

Today we are in the opposite situation. The exit door for carbon is being used more than the entrance, and thus the amount of CO_2 in the atmosphere is rising. This is not due to an increase in the Krebs cycle, however; it is occurring because of the actions of the human population. When we burn the remains of dinosaurs and extinct forests to power our cars or heat our homes, we are returning to the atmosphere the carbon that was stored underground around 100 million years ago. In effect, we are adding to the exit part of the cycle while the rate of entry remains steady.

What will happen as a result? Scientists differ on this; the consequences of such large-scale changes are difficult to predict with any degree of confidence. It is known, for example, that the oceans are large enough to absorb all of the excess carbon dioxide—it will simply diffuse into the water and settle out as calcium carbonate ($CaCO_3$)—but there is the problem of rate. If we are adding CO_2 at a faster pace than the ocean can absorb it, CO_2 will continue to accumulate in the atmosphere, as it has been since meteorologists began to keep accurate records in 1957 (Figure 3.10). And at some point, the warming effect of CO_2 will cause the average global temperature to rise.

If an increase in temperature occurs, and it is high enough to partially melt the polar icecaps, there will be many ramifications—oceans will rise, and some low lying countries like Bangladesh and Panama could find themselves under water. In the United States, coastal cities like Miami and New Orleans could be threatened. Another result could be that the major centers of agricultural productivity would shift northward—good news for Canada, bad news for the farmers of Iowa. I use the word

---

7. Chapter 4 explains how such precise regulation is achieved.

Figure 3.10 Recordings of the CO$_2$ content of the atmosphere from Mauna Loa, Hawaii, reveal the natural and human-induced changes in the carbon cycle. The annual fluctuations reveal natural, short-term adjustments in the carbon cycle. During the winter, respiration exceeds photosynthesis, so CO$_2$ is added to the atmosphere faster than it is taken up by plants. During the summer, the reverse is true. The steady, long-term increase in CO$_2$ is believed to be due to human activity. When we burn fossil fuels for energy, extra CO$_2$ is added to the atmosphere.

"could" because these are educated guesses, which we have to rely on in the absence of experience. The only thing we can say for certain is that the carbon cycle will eventually balance—left to its own devices, the earth takes care of itself. But how long will it take? That is the crucial question that we may regret to answer in the future.

The carbon cycle is somewhat like the cycle of the spirit, samsara, that is described in the early texts of Lamaism. After death, they say, one's spirit first resides invisibly in the sky, eventually becoming the air itself; then masses of spirits condense into clouds, and fall as rain. After soaking into the soil, spirits are absorbed into plants to be eaten by men. There, they become the life-giving gametes, and as such, are poised to begin life again.

Clearly, the founders of the faith recognized the cyclic nature of life, and gracefully linked supernatural events to the earthbound cycles of nature. The supernatural aspects of their beliefs extend beyond the reach of science, and thus are the material of religion and philosophy; but to the biologist, the cycles found in the scientific study of life are no less magnificent.

? Q & A with Jeff & Jen

JEFF: Very impressive stats on the ATP cycle, Jen. A cell has 500 million molecules, and uses them all up in a minute?

JEN: Those stats are for a muscle cell that's actively contracting, but it's in the ballpark for other cells.

JEFF: So if a cell doesn't recycle the ADP and P, it's a goner in no time. Does that ever happen?

JEN: Yes. As a matter of fact, anything that cuts off the oxygen supply to a mitochondrion will have that result. Oxygen is required for the electron transport system of a mitochondrion to work, and without the ETS, a mitochondrion can't complete the ATP cycle. That's why cyanide is such a potent toxin—it prevents the ETS from using oxygen, even if oxygen is available.

JEFF: And thus a minute later, the cell's ATP supply is gone, and the cell dies.

JEN: Right. There are a bunch of toxins that interfere with mitochondria. The point is, any interruption of the ATP cycle is potentially lethal. Cycles in general are very precise—interruptions in them lead to all kinds of problems.

JEFF: You also emphasized that they're linked. So does that mean that if one cycle is broken, the other cycles are affected?

JEN: You bet. Think about what happens during a heart attack. The basic problem is that the heart cycle stops. That alone isn't the main problem. The bad thing is that when the heart stops, the cycle of circulation does, too. Oxygen doesn't get to the tissues. Some tissues can tolerate that, but brain cells can't—they begin to die about six minutes after circulation has stopped.

JEFF: Hey, I just realized something. The circulation cycle is linked to the ATP cycle, right? Since oxygen isn't getting to the brain, the mitochondria in the brain cells can't complete the ATP cycle, so the brain cells die from lack of ATP.

JEN: Pretty good, Jeff. Now you're getting the picture of how these processes are mutually dependent.

JEFF: But I still don't understand what interrupts the heart cycle to begin with.

JEN: There are dozens of causes. But the most common one is that cardiac artery is blocked by a fat deposit. That's the artery that supplies oxygen to the heart muscle. So, the heart muscles themselves are deprived of oxygen, and they go into spasms. If the blockage is severe enough, it can cause some of the heart muscle to die.

JEFF: Part of the heart can die, and the person can live?

JEN: If it's only a small patch, about the size of a dime. So it doesn't prevent the heart as a whole from working. But it can still interfere with the timing of contraction. If there are a few surviving cells scattered through a dead patch, it can become something called a reentrant circuit. Basically it acts like a third pacemaker and sends out messages that desynchronize the heart.

JEFF: I get the point. Let's go on to the atmospheric cycle of carbon. What's the interruption in this one?

JEN: Well, it has to do with the amount of CO_2 in the atmosphere. There are two molecular cycles involved: (1) the Calvin cycle in photosynthesis, which removes CO_2 from the air, and the Krebs cycle, which adds CO_2. If the amount of CO_2 in the air is going to stay constant, the two cycles have to balance.

JEFF: And they don't?

JEN: Well, that's hard to say. Even if they do balance, the problem is that we are supplementing the Krebs cycle. By burning fossil fuels, we add CO_2 to the atmosphere. That's equivalent to speeding up the Krebs cycle while holding the Calvin cycle steady.

JEFF: And more CO_2 will mean higher temperatures.

JEN: Probably. It has been proven that the amount of CO_2 in the atmosphere is rising, but not that the average temperature on earth is rising.

JEFF: Okay, enough. I've got cycles on the brain by now. Just for good measure, can you give me some tips on reincarnation?

JEN: Funny, Jeff. When do you want to start?

RELATED RESOURCES

Surgical treatment of cardiac arrhythmias, A. Harken, *Scientific American* 269 (July), 68–74, 1993.

Tibetan tourist thangkas in the Kathmandu Valley, Y. Bentor, *Annals of Tourism Research* 20, 107–137, 1993.

Dangerous dance of the dividing cell, S. Young, *New Scientist* 134 (June), 23–27, 1992.

The lemming phenomenon, L. Hansson, *Natural History* (December), 38–43, 1989.

Warming up to hot new evidence, C. Moore, *International Wildlife* 27 (January/February), 21–25, 1997.

Predation and population cycles of small mammals, E. Korpimaki and C. Krebs, *BioScience* 46 (November), 755–764, 1996.

chapter 4

SPEED

Theme

Regulation is essential to every process in nature. Each must be turned on, modulated while it is on, and then turned off. If processes were not controlled, life at every level of The Living Staircase would collapse. Feedback regulation, when the end result of a process circles back to affect specific elements of the very process that created it, is a common and very effective regulation strategy.

Examples

- Muscle contraction
- Neurotransmission
- Glycolysis
- Hypothalamic function
- Transcription of genes

"*G*ood evening, ladies and gentlemen, this is *These Sporting Times*, and I'm Jeff Meghan coming to you live from our studios in Los Angeles. Today we have a special show for you that combines the best of what Speed and Science have to offer. We're going to be looking frame by frame at the fastest 100 meters ever run by a woman. In this exercise alone, there is nothing unusual, as sportscasters like myself have been analyzing this particular feat since the moment it occurred. What's different today is that we have a physiologist and trainer, special guest Pearl Garcia, to help us understand the biological regulation that is going on in the runners during the race. We have here a unique opportunity to relive history and learn biology simultaneously. Welcome, Pearl."

"Thanks, Jeff, I'm glad to be here, and I'm looking forward to getting started."

"Let's not waste any time then. Let me set the stage: This is the 100-meter dash—the race that traditionally determines the fastest man and woman on earth, a distinction so honored and respected that we use cutting-edge technology to determine who wins in what time: electronic sensors on the starting blocks, automatic cameras to catch photo-finishes, timing devices that can separate runners to within one-hundredth of a second.

"The first fifteen one-hundredths of a second is key, Pearl. As long as it takes you to blink or to think a thought. If the mind and body are not perfectly synchronized with the starter's gun, the race could be lost right there.

"In that first few hundredths of a second, Florence Griffith Joyner started a race that she would win 10.49 seconds later. A time so fast that experts had predicted that no woman on earth would approach it until well past the year 2000!

"I remember the race well, Jeff. Everyone, including the winner I think, was shocked at the record time. Now, we are about to accompany Florence Griffith Joyner, or FloJo as the country and the world soon came to know her, on that short journey. As we watch FloJo, we'll pause the videotape and ask ourselves some questions about regulation. Where does the energy she uses to run come from and how is it controlled? How does the human body, or that of any organism, regulate its functions? For example, why don't runners like FloJo, moving at 30 miles per hour, overheat? What starts the energy surge? How does such a process stop once it's started?"

"I must admit, Pearl, that I haven't really considered many of these questions much, but they must be important for runners to consider, since it's their bodies that are being tested. The answers to these questions are, of course, important to us, too, since everything that occurs in runners also occurs in us—perhaps a bit slower. Before we let the tape run and the race start, maybe you could fill us in a little on the basics of regulation, Pearl."

"Sure. Regulation is all around us. Your director, Jeff, regulates this television show, telling you when to break for commercials, when to return to the audience, and when the show is over. In a way, your sponsors regulate you, too. Regulation is simply the control and adjustment of a process.

"As FloJo and her competitors lined up for that world-record 100m Olympic trial in Indianapolis, regulation was at work. The starter and her gun induced the runners to start the race; the sprinters regulated their speed based on the nearness of their competitors, the distance they had traveled and had yet to travel, the sound of the crowd, and how much energy they had. The finish line, placed exactly 100m from the starting line, signaled FloJo and the others to stop."

"Hmm, I guess in some ways then, Pearl, that all those hours of training, coaching, and psychological preparation also regulated, at least indirectly, the sprinters during the race. Wouldn't you say?"

"Definitely, Jeff. You're getting the idea."

0–0.2 Seconds: The Gun Sounds, Muscles Move

"Okay. Let's get started. The runners are set. It's a fair start and . . ."

"Freeze it right there, Jeff. Okay, what just happened? FloJo's ears transformed the signal of the starter's gun to an electrical signal and sent it to her brain which translated it to a signal, which was sent down her spine to her neuromuscular junctions, the places where nerve and muscle meet. In order to pass from the nerves to the running muscles, the electrical signal sent from the brain had to have been translated into a chemical signal, a neurotransmitter.[1] This neurotransmitter then flowed across the neuron-muscle junction. On FloJo's running muscle cells, receptors bound to the transmitter. This caused the release of calcium stored deep in these muscles, and . . ."

"Whoa, Pearl, slow down. Let me get this straight: the signal of sound is translated to an electrical nerve signal, which is translated to a chemical nerve signal, which reaches the muscles and tells them to release calcium. This all happens in less than a second?!! What does calcium have to do with running, anyway?"

"Well, to answer your first question: yes, this clearly happens incredibly rapidly. To answer your second question: calcium is a major regulator of **muscle contraction**.

"Let me back up for a minute, though, Jeff. Muscle is composed of thick and thin filaments. The thick ones are made of proteins called myosin, and the thin ones of

---

1. See Chapter 9 for more details on the process of neurotransmission and signal transduction. See Chapter 8 for a discussion of energy conversion, which in this case is sound to electrical energy transformation.

another protein called actin (Figure 4.1). Contraction occurs when these filaments slide over each other. Can you guess what might happen, Jeff, if this sliding contraction is *not* carefully regulated?

"Uhh . . . I suppose if the relax signal isn't received, then muscle might get stuck in mid-contraction and not be able to relax, like a cramp or something. But, Pearl, what about the calcium; where does it fit in?"

"I'll get there, Jeff. Those were good guesses about what happens when the regulation of muscle contraction breaks down. Maybe you should consider another career. If you look at this picture here (Figure 4.1), you can see in more detail how contraction works—and don't worry, Jeff, we'll get back to the race in a minute."

"When the muscle receives the signal for action from the central nervous system, muscle cells release calcium from special storage centers. Calcium enables the actin and myosin filaments to bind each other, resulting in the release of energy (ATP, adenosine triphosphate) stored in the myosin. This binding causes the myosin, attached to actin, to bend, shifting the position of the myosin and the actin bound to it. The result is that the filaments slide over each other. The idea of proteins changing shape to regulate processes is a common one in biology.[2] The sequential bending and sliding, occurring hundreds of thousands of times in every muscle, results in contraction."

"Huh, sounds sort of like a sling shot mechanism."

"Sort of, Jeff. I think it's better, though, to compare contraction to the cocking and firing of a starter's pistol, like the one we just heard start FloJo's race. Muscles use a 'protein gun' to regulate filament sliding (Figure 4.1). First the gun is cocked; this is equivalent to the myosin binding with actin in the presence of adenosine diphosphate (ADP) and inorganic phosphate (P_i). Then, like firing the gun, the ADP and P_i are released and the myosin and actin slide over each other."

"Okay, Pearl, but all we've done is get the sprinters' muscles contracting. How do they stop contracting and relax? Like we just agreed, if FloJo's muscles had only gotten her this far, they would seize up."

"Good point. Most regulation occurs in on/off cycles.[3] Running and, in fact, all other movement result from a combination of opposing muscles contracting and relaxing. Once the brain stops sending the stimulatory electrical pulse, the nerve cells stop releasing neurotransmitter across the neuromuscular junction, and the muscles

---

2. See discussion of how protein conformation is important in Chapter 1 (hemoglobin) and in Chapter 9 (cell-surface receptors).
3. Many metabolic processes are cyclic; for information about the characteristics and importance of cycles, see Chapter 3.

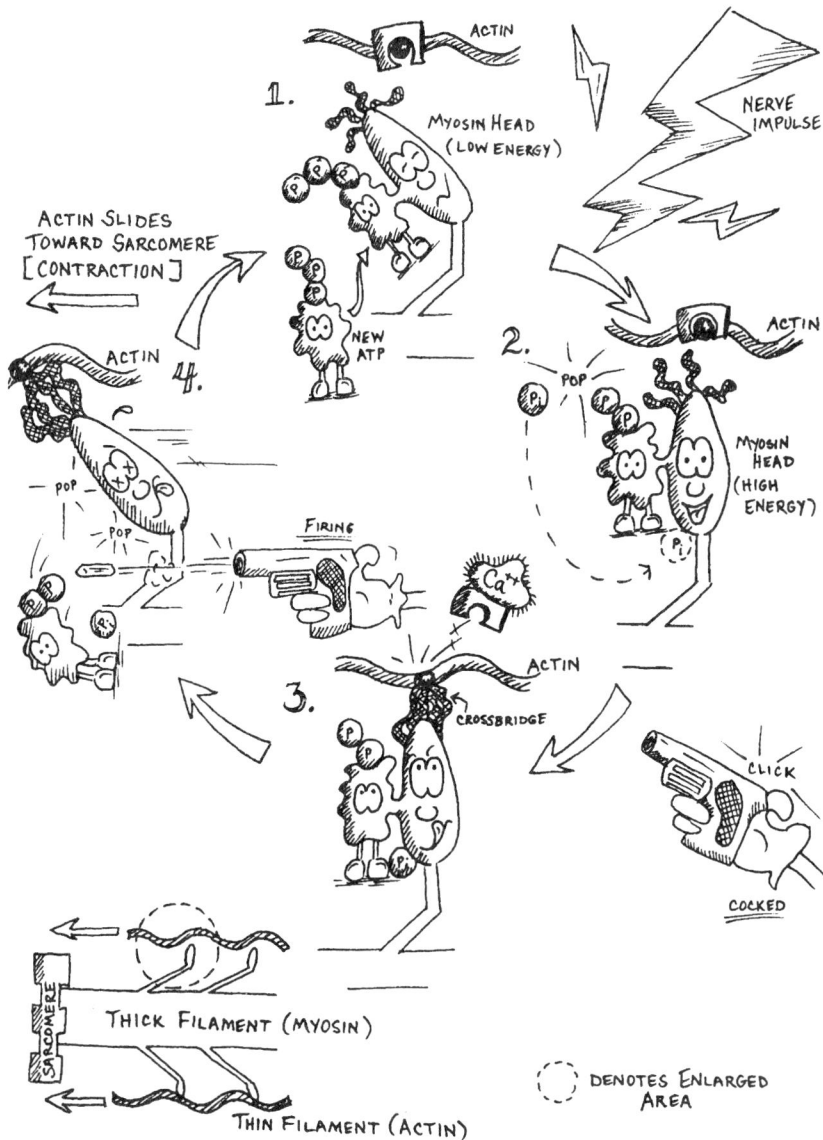

Figure 4.1 Muscle contraction. The cycle shown is a magnification of the circled section of muscle in the lower left. (1) Myosin is bound to adenosine triphosphate (ATP) and is in its "low-energy" state. (2) Myosin attains its high-energy state once signaled to do so by a nerve impulse, hydrolyzing its ATP to adenosine diphosphate (ADP) and inorganic phosphate (P_i). The nerve signal also releases calcium from storage in the cells. The calcium interacts with the proteins allowing (3) myosin to bind to actin, and the system is "cocked." Then (4) the excitatory signal is no longer sent, calcium is removed, and the system returns to its low-energy state, sliding (or "firing") the actin in the process. (5) New ATP binds the myosin and it is released from actin.

no longer release fresh calcium. The 'reloading' of the system occurs when ATP again binds to myosin, myosin no longer binds with actin, and the system is relaxed and ready to contract again."

"I'm wondering about your 'fresh calcium,' Pearl. I mean what happened to all the calcium that was released in the muscles to start with? Does it just disappear?"

"I'm really impressed with your insight, Jeff. Our audience must think I've given you cue cards the way you're asking all these astute questions."

"Well, thanks, Pearl. Go on, please."

"Okay. Let's see—two energy-requiring calcium pumps and special calcium-binding 'vacuum cleaners' return the released calcium to storage to help the system return to its resting state. Interestingly then, both contraction and relaxation require energy. Regulation of muscle tissue is accomplished by regulation *of* molecules, as well as by regulation *by* molecules."

"As you alluded to earlier, Jeff, an incredible thing about all this regulating, which began back with the starter's gun being shot, is that it occurs in less than two-hundredths of a second. As we mentioned earlier, the start of a race as short as this one is extremely important. It is not unusual for runners to finish within a hundredth of a second of each other, so any advantage gained at the start is crucial."

"Thank you, Pearl. We'll continue with the rest of the race and our scientific glimpse into history after this message from our sponsors."

0.2–6 Seconds: FloJo Is Flying and So Is Her Metabolism

"Welcome back to our show, folks. I'm here with noted physiologist Pearl Garcia, and we've been analyzing the famous world-record dash of Florence Griffith Joyner."

"Pearl, as we watch the next few seconds of the race in slow motion, one can't help but be amazed by the fluidity and grace of these runners. FloJo looks like she's not even touching the ground! Where does she get the energy to run like that?"

"I'll start with metabolism, Jeff. This is a word we use to describe all the processes in our bodies that make and use energy. Proteins called enzymes are the great regulators of metabolism.

"Another central concept to understand before we get down to the nitty gritty is feedback regulation. The products of any metabolic process almost always feed back to regulate the very systems that helped produce them. Are you with me?"

"I think so, but where does the energy for metabolism itself come from in the first place . . . I guess that's where food comes in, huh?"

"Yes, it's the carbohydrates, fats and proteins in food that the enzymes metabolize, or break down, to produce energy. I think once we get into some examples it will become clearer.

"Remember that initial huge thrust of energy FloJo used to start the race? FloJo used the major energy source, ATP, stored in her muscles for that first explosive muscle contraction (Figure 4.1). But she only had enough ATP stored to last for about one second of sprint."

"Excuse me, Pearl, I remember this part from high school: mitochondria made the ATP, right?"

"Well, yes and no, Jeff. Normally, the process called cellular respiration, which occurs in mitochondria like you say, provides ATP;[4] but this process is not fast enough for FloJo's immediate energy needs (Figure 4.2). Instead, the transfer of a high-energy phosphate group from a compound called creatine phosphate to ADP produces ATP and provides the next available source of energy (Figure 4.2). Think of each of the phosphate groups in these processes as carrying potential energy."

"Huh, I guess I slept through that part of biology class; but, I won't let you sneak by with a bunch of fancy terminology—how does this 'high-energy phosphate group' transfer happen? Something must regulate the transfer; otherwise it would occur all the time, wasting energy like crazy."

"Bingo. An enzyme called creatine kinase regulates the creatine phosphate reaction (Figure 4.2). A kinase is an enzyme that transfers phosphate groups between molecules. Creatine kinase transfers a phosphate from creatine phosphate to ADP. The breakdown of creatine phosphate allows for the instant replacement of the phosphate lost from ATP on the myosin during contraction (Figure 4.1).

"However, after about six seconds of hard sprint, right at the point we freeze-framed the race a minute ago, FloJo's muscles have run out of even creatine phosphate. Before we discuss her next source of energy, Jeff, why don't we continue with the race tape."

6–10 Seconds: More Energy Quick—pH and PFK

"At the six-second mark FloJo, in a graceful burst of speed, breaks away from the pack. For a race so short, she already has a substantial lead! So, Pearl, I suppose the next question is: if stored ATP, and then stored creatine phosphate have both run out, where do FloJo and the others get the energy for the last four seconds plus of the race?"

"**Glycolysis.** Glycolysis, like the name implies, 'glyco' meaning sugar and 'lysis' meaning breaking up, is a process that metabolizes sugar to produce ATP. It does not

---

4. See Chapter 8 for a description of how mitochondria synthesize ATP.

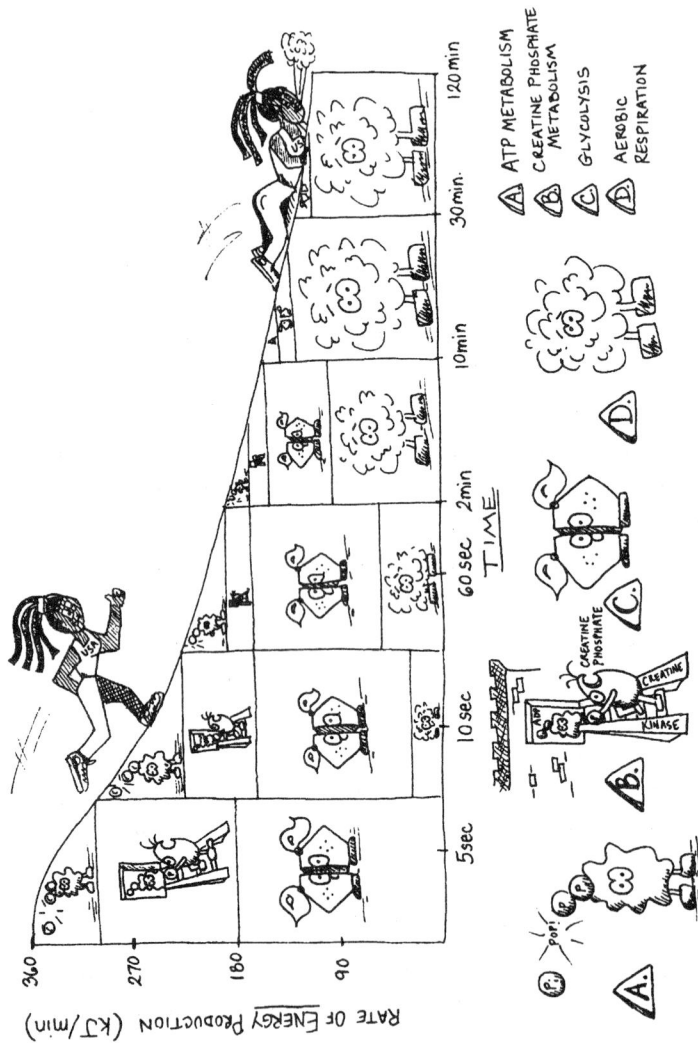

Figure 4.2 Metabolic sources of energy during intensive exercise. This graph shows where our energy comes from when we're running full out. For example, for the first six seconds of the race (the first "bar" in the graph), about 60% of FloJo's energy comes from glycolysis and about 40% from creatine phosphate breakdown. Note that the input of different energy sources varies with time, and that the total maximum energy output possible decreases quickly with time. The regulation of each type of energy metabolism is intimately associated with the regulation of the others. Of course, FloJo was only worried about the first 10 seconds of this graph. Below the graph are depicted the icons for the different energy sources. "A" shows ATP broken down into ADP and P_i. ATP is our major source of free energy. FloJo was using ATP at the rate of about 0.2 kilograms per minute in her race! ADP can also be further metabolized to adenosine monophosphate (AMP) and P_i. "B" shows the enzyme creatine kinase catalyzing the transfer of the phosphate group of creatine phosphate to ADP to produce energy for the first few seconds of intense exercise. "C" symbolizes glycolysis, the "splitting" of glucose detailed in Figure 4.3, and "D" represents aerobic respiration.

require oxygen, and neither does the creatine phosphate reaction I mentioned earlier. Glycolysis, which occurs in the cytoplasm of cells (mostly muscle cells in this case), is the major oxygen-independent or anaerobic producer of energy. As I said, we also produce energy by cellular respiration; a process that *does* require oxygen (aerobic). In all three cases, it is the addition of one of those high-energy phosphate groups to ADP that produces energy in the form of ATP."

"So you mean, the ATP made in mitochondria that I learned about is never even used in FloJo's sprint?"

"That's right, Jeff, not significantly anyway. But, don't worry: cellular respiration *is* the major energy producer in us. Although glycolysis makes energy available faster than your mitochondria can, as you can see in Figure 4.3, it yields a paltry net of two ATP's per glucose molecule, as compared to 36 ATP's per molecule made during respiration (Figure 4.3). Cellular respiration just can't cope with huge, fast bursts of energy. We have these special energy-producing mechanisms to deal with those. In longer races and in day-to-day life, though, mitochondria are it."

"I'm crushed. Oh, well. How does the glucose, say in a candy bar, wind up being available for glycolysis in muscle cells? A lot must happen between chewing and ATP, no?"

"Quite a bit, Jeff. More than we have time for in the show. Suffice it to say that many other groups of enzymes digest your candy bar into glucose, among other products. Yet more enzymes then string the glucose molecules together into long chains called glycogen. It is the glycogen that is stored in the sprinter's muscle cells and is quickly converted back to glucose for glycolysis to use."

"I noticed, Pearl, that in your picture illustrating glycolysis, you have one enzyme circled, phosphosomething-or-other. What does this enzyme do that makes it so special?"

"Jeff, in any metabolic process, one or two key enzymes are not only necessary for that process, but in addition are sort of like control centers for the entire process. Phosphofructokinase (PFK) is the control center for glycolysis. The name shouldn't scare you. Enzymes are named for the substrate they act on and the job they do. So, 'phospho' and 'fructo' just mean the enzyme works on a molecule that has a phosphate group, which we've already discussed, attached to fructose, a type of sugar (Figure 4.3). Finally, the 'kinase' part means, as we mentioned, that PFK moves the phosphate group from or to its substrate, the fructose.

"Think of an enzyme like PFK as a control center for running shoe production in the United States; sprinting shoes are the product resulting from this enzyme's work, and the organism is the running shoe industry. If there are too many sprinting shoes on the market, the control center stops producing them. This halt in production affects many other companies. So leather, metal (for spikes), and nylon suppliers and the

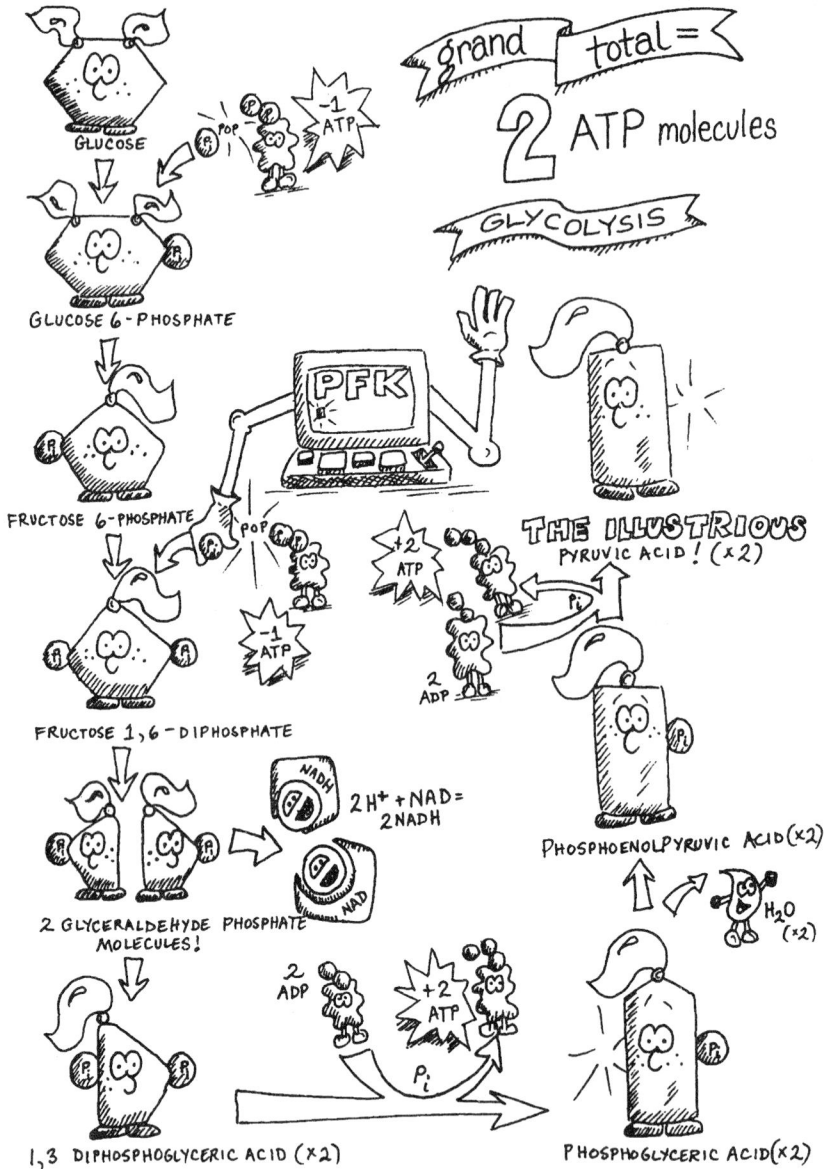

Figure 4.3 Glycolysis; this is the major source of energy for FloJo. The diagram shows the regulatory enzymes for each step, but we only talk about phosphofructokinase (PFK), the key controller enzyme of glycolysis, in the text. Two ATP's are used early, and then four ATP's are produced later, resulting in a net of two ATP's for each glucose molecule metabolized. Note the hydrogen ions that glycolysis produces. These also have a role in regulating the pathway. Glycolysis occurs in the cytoplasm, while cellular respiration takes place in the mitochondria as discussed in Chapter 8.

companies transporting these supplies also decrease their activity. And because there might be too many shoes produced for several different reasons—for example, a miscalculation by marketers or an overseas competitor or imitator making too many similar shoes—the control center must be able to respond to several different types of inhibitory signals."

"Quite a mouthful, Pearl. I think the main point, though, is that PFK is the energy production control center for glycolysis—right? So, according to what you said earlier, products of glycolysis must feed back to regulate PFK, just like a glut of running shoes feeds back to regulate production at the factory."

"Good, Jeff. For now, FloJo wants only more energy; she is, after all still in the middle of the race. So she needs more ATP, more glycolysis, more PFK activity. So, let's first look at *stimulators* of PFK (Figure 4.3). Since you're doing so well, what would you guess are some factors that stimulate PFK and, therefore, glycolysis?"

"Uh, well, let's see. Glycolysis makes ATP by using PFK . . . no, it makes ATP by adding phosphate groups to ADP, right? When no phosphates are being added to ADP, there's a lot of bored ADP waiting around. Also, since a lot of ATP on the myosin in FloJo's muscles has already been used, resulting in a release of ADP and P_i (Figure 4.1), even more 'free' ADP is present. More free ADP around is a sign that more A*T*P is needed, no? What if ADP stimulated PFK to increase glycolysis; this would mean more phosphates would be added to ADP itself, resulting in more ATP, right? That's it; that's my guess: ADP stimulates PFK."

"Very nice, Jeff. I'm beginning to think I should be interviewing you and not vice versa. When FloJo's muscles contract, ATP breaks down to ADP like in my contraction picture (Figure 4.1). This ADP, as you guessed, then stimulates PFK, which in turn stimulates glycolysis, which metabolizes the glucose stored in muscle glycogen, producing more energy.

"We have enough energy stored in glycogen to easily sprint full-out for 80 seconds or more; however, most of us can only sprint at full speed for about 20 seconds. *Negative* feedback prevents our muscles from using all their stored glycogen, and, as we'll see shortly, it's a good thing."

"Pearl, if I was right about ADP stimulating PFK, then it would follow that ATP, or too much of it anyway, inhibits PFK (Figure 4.3A). As we were discussing, we don't want energy production to continue forever, but for only as long as we need it."

"Yes, ATP is a negative feedback inhibitor of PFK and glycolysis, just as you suggest. Hydrogen ions (H^+) are another inhibitor of PFK. They are a byproduct of glycolysis.

"If FloJo burned all her glycogen, she would collapse before she finished the race. H^+'s are a byproduct of the whole energy-producing process. Therefore, if glycolysis consumed all glycogen (glucose) present in muscle, it would significantly increase the level of H^+ in the cells, right? To measure the level of H^+ (acidity), we use the pH system: the greater the concentration of H^+, the lower the pH; therefore, the more acidic is a given solution. In most of our cells, a pH of 7 is normal. A pH below 6.2 in most cells, including muscle, is lethal (we tolerate pH levels as low as 4 in our digestive system, but that's another story). If we metabolized all the muscle glycogen we had available, the result would be a pH of 5 in our muscles and possibly death!"

"Wow, Pearl, since FloJo's muscle cells have had glycolysis cranked up for eight seconds now, pumping out hydrogen ions, it's a good thing for feedback inhibition or the pH in her muscle cells would be dropping to dangerously low levels."

"True, Jeff, and one intriguing result of this is a phenomenon we're all familiar with: the farther we continuously run, the slower we are capable of running. Why? Well, the reason is that the more exercise your muscles do, the more H^+ are produced, the less efficiently the muscles work, and the slower you run. But instead of continuing to use all the glucose available—a potential disaster—PFK is inhibited by the high level of hydrogen ions."

"I'll finish the loop, Pearl. When PFK activity is turned down, glycolysis makes less ATP and less of the dangerous H^+ ions."

"You've got it, Jeff. I've summarized PFK regulation in this figure." (See Figure 4.3.)

"Y'know, Pearl, during this whole conversation, I've been sort of puzzled. I think I've got the idea of feedback inhibition down pretty well, between the shoe factory analogy and the PFK example, but I wonder . . . you keep saying that this 'inhibits' or that 'stimulates' PFK. What does that really mean? In the analogy of the running shoes, people stop buying, so the stores stop ordering, so the factories stop making when the managers halt their shoe-making workers and machines. But, how do molecules like ions and ATP actually regulate enzymes?"

"Well, Jeff, without getting into it too heavily, because we want to get back to the race, I'll just say most 'control center enzymes' like PFK are made up of many components and thus have a complex three-dimensional structure.[5] All enzymes have an active site with which their substrate interacts, and PFK and other control center enzymes are no exception. Remember we talked about how PFK is named for its substrate, phosphofructose?

"Well, inhibitors and stimulators actually physically interact with the active site and change its shape to either enhance or block the substrate's interaction with the

5. See Chapter 1 for a discussion of protein structure.

enzyme. An analogous inhibitor of an air-traffic control tower would scramble the radio signal between the tower and planes, while an activator might add more controllers to the tower. Understand?"

"Very clearly, Pearl. Thanks. Folks, as we go out for a commercial, let's watch the first 10 seconds of FloJo's record-shattering race again. We'll be right back. Don't run off. . . ."

Hypothalamic Thermostats and Anabolic Steroids

"Welcome back, folks. I'm Jeff Meghan and I hope you've been enjoying our scientific analysis of the fastest race ever run by a woman—courtesy of Pearl Garcia, physiologist and track and field maven.

"You know, Pearl, in watching the race one more time, and in thinking about the amounts of energy FloJo is expending here, I can't help wondering how her body handles all the heat she's generating. We must have some sort of thermostat system in our bodies that keeps us from meltdown, no?"

"Yes, we do, Jeff. In fact, why don't we discuss overheating from two angles. First, we'll talk about our built-in thermostat, and then, if we have time, we'll touch on how the proteins in our body deal with sudden heat stress.

"As you say, FloJo and the others are creating a tremendous amount of heat in their muscles. The thermostat that controls our temperature is located in a part of the brain called the hypothalamus. This organ is a mediator between the nervous system and the endocrine system."

"Uh, sorry Pearl, what's that last thing?"

"The endocrine system includes the group of organs that secrete hormones. Hormones are the great regulators of our bodies and other animals'.[6] They move through the blood and communicate with various cells and organs throughout the body. We'll discuss some examples of this in a minute.

"First, let's define the hypothalamus. It is a major control tower of regulation. It maintains our physiological state in balance with the environment and, therefore, is a center of feedback control. Information enters the hypothalamus from all over the body by way of hormones and the nervous system. It sorts the data and responds; its commands are also in the form of hormones.

"A body temperature increase of as little as 0.01 °C causes the hypothalamus to hormonally stimulate the sweat glands in the skin to secrete perspiration, moistening the skin. The evaporation of the perspiration helps cool the body.

---

6. See Chapter 9 for a more detailed discussion of hormones.

"And, Jeff, it's important for us and our viewers to realize that the hypothalamus also contains centers for controlling hunger and thirst and plays a role in regulation of sexual development, mating behaviors, aggression,[7] pleasure, and growth."

"I see. Pearl, this is ringing a couple of bells from some stories I've covered and read recently. Is this the part of our regulatory system with which anabolic steroids interact, by any chance? Remember, it was the very Olympics that FloJo was qualifying for in this race that saw Ben Johnson stripped of his 100m gold medal. He was suspended from track and field for testing positive for an illegal anabolic steroid. Does the hypothalamus have anything to do with all this?"

"Yes it does, Jeff. Steroids do affect regulation, but if you want to get into the problems of drugs, it's essential that our audience first understands two facts: (1) steroids are a class of hormones that our bodies produce *normally* which endocrine glands like the hypothalamus use to communicate with the rest of the body and (2) scientists around the world are engaged in extensive research to identify steroid drugs that we and animals can use to *positively* affect many diseases or other problems."

"I know what you're saying, Pearl; steroid drugs have a bad name. Can you give us some examples of the 'good' uses of such drugs?"

"Sure. Doctors prescribe steroids to aid people who have steroid deficiencies. One therapeutic use of steroids is the prescription of growth hormone for people with growth problems. Steroids can also aid in the healing of injuries. Many of us have taken the steroid prenosone, which reduces swelling of, for example, the lungs in asthmatics. The vet has prescribed it for my dog to stop him from scratching his flea bites. For obvious economic reasons, animal geneticists have also treated cattle with growth hormone and other hormones to try and produce larger animals with, say, more meat and less fat. Doctors even prescribe synthetic testosterone, like the type Ben Johnson tested positive for, to aid older men gain back strength, sexual desire, and better spatial cognition and word memory."

"It's incredible, Pearl, how scientists have learned to harness or mimic nature to help society. But, I guess like most things there's a flip side to this good part, and, unfortunately, the bad stuff gets all the headlines—like the Ben Johnson fiasco. We're sort of getting away from FloJo's magnificent race for a few minutes, but I think this issue of drug abuse is related to sports and society in general and that people should really hear as many of the facts as they can about this stuff. Do scientists understand yet how steroids can make us 'superhuman'?"

"Partially, Jeff; the effects of taking these strength-enhancing drugs, specifically called anabolic steroids, vary from person to person depending on sex, age, dosage,

---

7. See examples of the behaviors and development these hormones regulate in Chapter 9.

and many other factors. As we'll see, the hypothalamus sits atop a vastly complex web of regulation affecting many aspects of behavior and development. I guess what I'm saying is we understand a little, but far from everything.

"I'll give some background here first. All anabolic steroids are derivatives of the hormone testosterone. Testosterone is the major androgenic or 'male-making' hormone secreted by the gonads. Estrogen is the 'female-making' equivalent. However, both males and females have testosterone *and* estrogen in their systems. It is actually the ratio of the two that determines whether you develop as a male or as a female."

"Excuse me, Pearl, but this must be one of those areas where steroids can be helpful. If a newborn is having problems, for some reason developing as one sex or the other, can the child be given estrogen to make it female or testosterone to make it male?"

"That's correct, Jeff. The normal ebb and flow of hormones provides some nice examples of regulation (Figure 4.4). For example, in males, the hypothalamus secretes a hormone that regulates the release of luteinizing hormone (LH) and follicle-stimulating hormone (FSH) from the pituitary gland. LH in turn regulates the production of androgens, such as testosterone, by the testes, and FSH induces sperm production."

"Hold on, Pearl—even without looking at the figure, I can smell some feedback regulation going on here, just like in the PFK and glycolysis thing."

"Good, Jeff. Classic feedback loops regulate the levels of androgens and estrogens in the body. Negative feedback of androgens regulates the cycle at three different points in this process: it controls the concentrations of LH and FSH and the levels of the hypothalamic hormone that regulates them. LH and FSH themselves also feed back to inhibit the hypothalamic hormone that regulates them! How's that for confusing?"

"Not so bad, Pearl, but why so much feedbacking?"

"Hard to say for sure, Jeff, but our best guess is that the more levels of feedback regulation you have, the more you can fine-tune the process."

"It makes one wonder, doesn't it, Pearl? It seems like taking artificial anabolic steroids into such a finely tuned system of regulation is like dropping a stone in a lake—the ripples would just keep spreading out, some good, but . . . so why do people keep using such drugs?"

"Well, we haven't even got to the worst of it yet, Jeff, but the thing is: anabolic steroids work. Witness Johnson's incredible run. Together with intense weight-training and exercise, anabolic steroids can give you as much as 25 pounds of muscle in the space of ten weeks!"

"How in the world . . .?!"

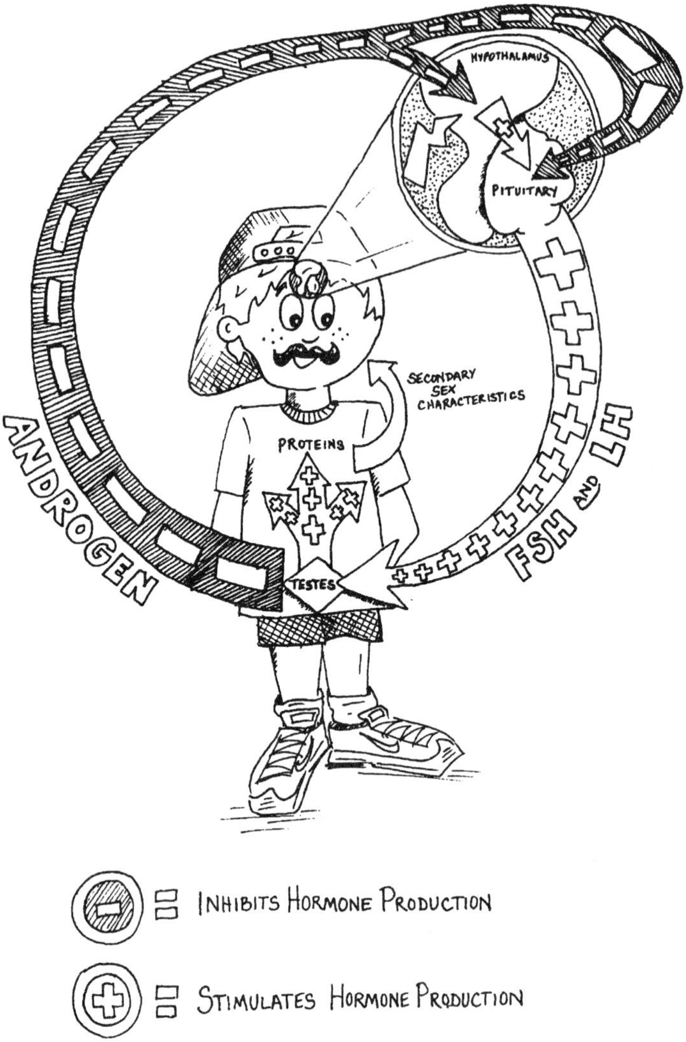

Figure 4.4 The hypothalamus and regulation, an example from the male. This is a diagram of some of the feedback loops that the hypothalamus regulates. The regulation involved in testosterone production is shown in more detail. Stimuli from other areas of the brain, and stimuli in the form of proteins produced elsewhere in the body (see Chapter 9 for an example, the hormones involved in sexual maturation) signal the hypothalamus. Hormones secreted by the hypothalamus stimulate the anterior pituitary gland to secrete follicle stimulating hormone (FSH) and luteinizing hormone (LH). These hormones, in turn, stimulate androgen production and spermatogenesis in the testes. Androgen feeds back to inhibit hormone secretion of both the hypothalamus and pituitary gland. Androgen also stimulates genes in the cells of the rest of the body to produce the proteins involved in creating the male secondary sex characteristics. It is in this process that anabolic steroids initiate their effects, but, as you can see, the regulation of other processes will be affected soon after.

"We're not exactly sure. The drugs seem to reverse the effects of muscle break-down during exercise and to increase the level of nitrogen stores in the body. When we talked about muscle energy, we didn't mention that during exercise endocrine glands called the adrenals secrete hormones that cause the breakdown of muscle tissue. Somehow excess testosterone in the form of anabolic steroids inhibits this built-in regulatory mechanism. So the muscle that would normally break down remains intact.

"Also, our bodies have stores of nitrogen. This nitrogen is essential for the production of amino acids, the building blocks of proteins, like PFK, actin, and myosin, and some of the steroids that we discussed earlier. Steroids cause an increase in the storage of nitrogen, which leads to the production of more amino acids and, therefore, more proteins such as actin and myosin that compose muscle (Figure 4.1). By these mechanisms, and probably others we haven't figured out yet, anabolic steroids used with exercise, cause an increase in muscle mass."

"But, Pearl, before we go to a commercial, I remember you said a while back that the hypothalamus, which regulates our normal steroids, also controls the hunger, pleasure and aggression responses. How does ingesting anabolic steroids affect the regulation of these things?"

"Well, taking artificial testosterone also leads to a desire to eat more and to exercise more and harder. Taking anabolic steroids stops the hypothalamus from signaling for the secretion of more testosterone, because the drugs fool the body into thinking it has already made enough testosterone. So, the body keeps building muscle tissue, when ordinarily it would stop.

"Some side-effects of steroids make sense to us, and some don't. Anabolic steroids knock askew the crucial sex-determining ratio of real testosterone to estrogen. Women on anabolic steroids tend to grow facial hair and their voices deepen. These drugs disrupt their menstrual cycles. Men grow breasts, their testicles shrink, and their sperm production decreases. Men on steroids often become impotent. Steroid use can also lead to heart, brain, and kidney damage, as well as high blood pressure, and a premature increase of cholesterol in the blood.

"Don't forget, Jeff, that the hypothalamus and all its connecting feedback loops also regulate physical development and behavior. Steroids can cause even more serious problems, such as stunting growth, when misused by young people who are still growing. Finally, taking steroids also can increase aggression, cause mood swings, psychoses, and lead to depression."

"Whew. On that fascinating, but disturbing note, we'll take a break, and be right back with *These Sporting Times* right after this word."

The Shock Factor

"We're back on *These Sporting Times*. We, myself your host Jeff Meghan, and physiologist, Pearl Garcia, have been looking at the biological regulation in the bodies of heavily exercising people, specifically sprinters, and especially Florence Griffith Joyner in her world-record 100m in the Olympic Trials in 1988.

"If we may move onto a new topic, Pearl. When we started examining the sprinter's regulatory response to excessive heat, you mentioned the hypothalamic thermostat, but then you also mentioned you'd like to discuss the body's response to heat stress from the angle of proteins. What exactly did you mean?"

"Well, Jeff, as you may know or have gathered from our talk so far, proteins are responsible for a huge number of jobs in us and all organisms. First of all, nearly all enzymes are proteins—so, few, if any, metabolic processes would occur in the body without proteins. Proteins are also structural; they compose the majority of our skeletons, our muscle, the skeletons of cells, our skin, our hair. Some hormones are proteins. The list goes on; my point is that proteins are absolutely essential. Not surprisingly, most of our proteins work best at 37 °C, our body temperature at rest. And they are very sensitive to any increase in temperature over 37 °C."

"So, Pearl, what happens to proteins when our body temperature suddenly increases before our thermostats can settle us back down to 37 °C? Do they melt?"

"No. Thanks to the so-called 'heat-shock response,' the proteins' structures are maintained. The molecules involved in this response and its regulation are similar in organisms as diverse as bears, barley, bacteria, and butterflies."[8]

"You mean to tell me, Pearl, that our proteins respond to extreme heat or stress just like *plant* proteins do?"

"That's right, Jeff. Before I get to the details of the heat shock response in the sprinters, though, I need to review some basics about gene regulation, because an understanding of it is necessary to fully appreciate the heat-shock response.

"Genes are made of deoxyribonucleic acid (DNA) and are located on the chromosomes in the nucleus of our cells. Genes encode the proteins that will actually do all the work we've been discussing. The genes are like a set of instructions for the cells. Two processes connect the DNA code and the resulting protein (Figure 4.5). One of these, transcription, converts the DNA information into a messenger ribonucleic acid (mRNA) molecule. The mRNA leaves the nucleus to carry the message to the

---

8. Evolutionary similarity is known as homology. Homology and its important implications are discussed in Chapter 6.

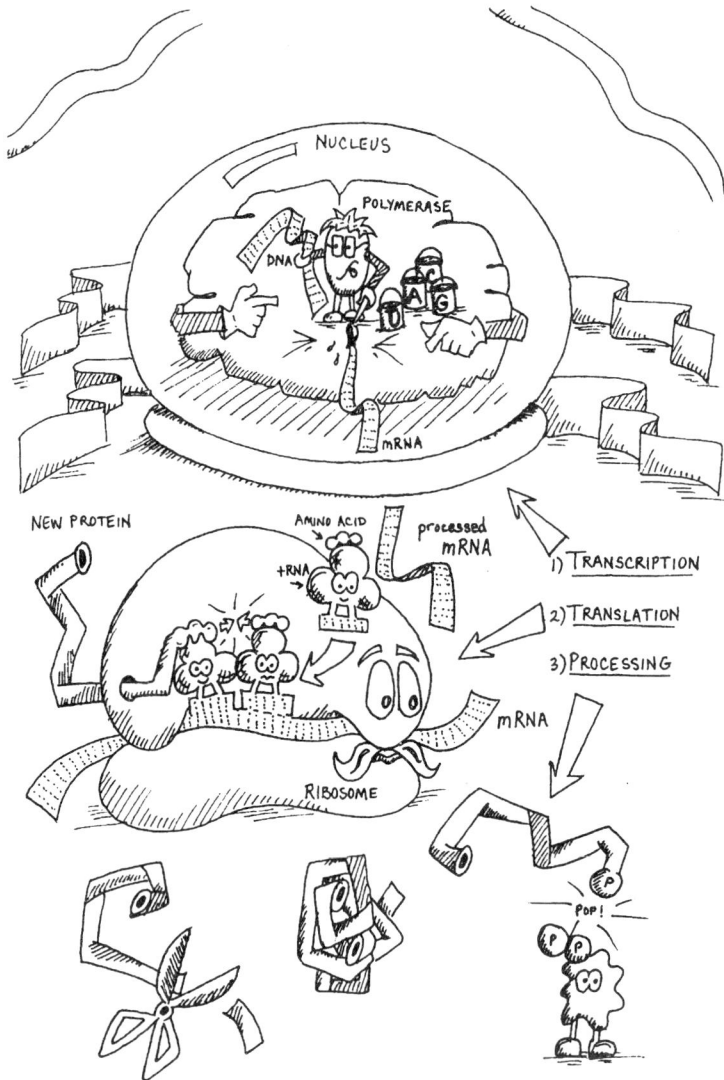

Figure 4.5 The protein production pathway in eukaryotic cells. Transcription occurs in the nucleus and involves general and specific protein factors, and ribonucleotides, the building blocks of mRNA. The polymerase reads the information encoded in the DNA and thus produces the messenger molecule, mRNA. The mRNA is processed by several enzymes before it leaves the nucleus and moves to the cytoplasm, where translation, the process of producing the protein encoded by the mRNA, occurs at the ribosomes. Translation also involves another kind of RNA, called transfer RNA, and enzyme machinery that aligns the corresponding amino acids encoded in the mRNA to produce a protein. Following translation, the protein may also be altered in numerous ways before it is functional. An example is the ATP that binds to and affects the function of the protein myosin in muscle contraction (Figure 4.1).

ribosomes in the cytoplasm. The second process, translation, occurs at the ribosomes and, using amino acids, converts the mRNA message into proteins."[9]

"Hmm, yeah, I remember the basic DNA-to-RNA-to-protein thing from biology class. But, Pearl, I can't say I remember hearing much about the regulation of these processes."

"Yes, Jeff, as with all biological processes, transcription and translation are regulated. For our purposes, we will concern ourselves only with the first process, transcription. The transcriptional regulation of a gene involves DNA sequences called promoters, because they are involved in promoting (and inhibiting) transcription. They are usually nearby the protein-encoding DNA sequences. Are you following all this, Jeff?"

"No problem so far. This is all news to me. How do the promoters of DNA regulate the part of the DNA that encodes proteins? What's special about them?"

"These promoters have a certain sequence to which proteins bind . . ."

"Hold on, *more* proteins??"

"Yes, I told you they do just about everything in the cell. Some proteins bind the promoter and help turn on transcription. Others bind it to inhibit transcription. It depends if the cell 'needs' the particular protein in question. This will become clearer in the heat shock example I'll discuss in a minute."

"So, Pearl, does each gene have its own unique transcription factors to turn it on and off? That would make for a lot of factors, wouldn't it?"

"Ah, good question, Jeff, and not an easy one to answer. The answer depends on the gene; some factors are specific to one gene, other factors interact with a certain subset of genes, and still other factors interact with nearly all genes. In fact, regardless of what other factors they may use, a certain group of general transcriptional factors are involved in turning on all of our genes. The most important protein of these general ones is RNA polymerase. RNA polymerase is the enzyme that . . ."

"I know, Pearl, enzymes are named for their substrate and what they do, like you said. So, RNA polymerase's substrate must be RNA, and DNA too, I guess, since it is involved in transcription, reading DNA to make RNA. Also, the polymerase makes a whole string, or polymer, of RNA. Thus, the name."

"There you go again, Jeff. Very good. Now, I think we're finally ready to talk about the cycle of gene regulation that controls the heat shock response in the sprinters' muscles.

---

9. See Chapter 3 for a more detailed discussion of translation.

"We have three players in this game. Follow them closely in my chart here (Figure 4.6). First are the Heat Shock Proteins (HSPs) themselves; under heat stress, they interact with the other proteins in cells to help them obtain and maintain their proper shape, and to help them move as they normally would within the cells. A protein must maintain its proper shape in order to perform its function. At normal body temperatures, most HSPs aren't needed and are present only at low levels. In response to high temperatures, however, their levels increase 10–100 fold.

"Second are the Shock Factors. They are proteins that regulate the HSPs. Before the sprinters build up intense heat in their muscles, the Shock Factors bind to the HSPs (1 in Figure 4.6) to prevent them from doing their job of stabilizing the body's proteins. Shock Factors inhibit HSPs just like hydrogen ions or ADP inhibit PFK in glycolysis. This is fine, since the body's proteins are not at high temperature and, therefore, are not in danger of melting, that is, losing their proper shape. When heat builds up, the Shock Factors change shape and let go of the HSPs. So, the two types of protein, HSP and Shock Factor, are free to do other things.

"We know what HSP does; it binds to and stabilizes the structure of proteins (6 in Figure 4.6). But, what do the Shock Factors do now? Well, they act to turn on transcription of the third player in our drama, the gene that *encodes* HSP (3 in Figure 4.6), by binding to its promoter. Yes, Shock Factors are also transcription factor proteins that turn on HSP transcription."

"Let me see if I've got this right, Pearl. Our bodies heat up and the heat releases the Shock Factor from HSPs. Then the Shock Factors bind to the HSP promoter? At that point, RNA polymerase must come along and make HSP mRNA that goes to the cytoplasm and is translated into HSP itself. This is just what the cell needs because HSP is the stuff that stabilizes proteins under heat stress, so the more HSP the better, right?"

"Right. Now, what happens to complete the heat-shock cycle of regulation? I mean, what happens after the hypothalamic thermostat has kicked in and the cells cool back down?"

"Well, without heat I guess the whole thing returns to what you might call 'normal,' no?"

"Right—there's no more heat stress to keep HSP and Shock Factor apart, and there's no longer any need for HSP; so, the two proteins, HSP and Shock Factor, can interact again, preventing each other from doing their respective jobs. Shock Factor doesn't bind the HSP gene promoter any longer, so no more HSP is made. Like you say, the cycle returns to its regular old 37 °C state. A crisis has been averted."

"Absolutely incredible, Pearl. Amazing to think of all the different interconnected, finely regulated processes occurring in us simultaneously, operating on every

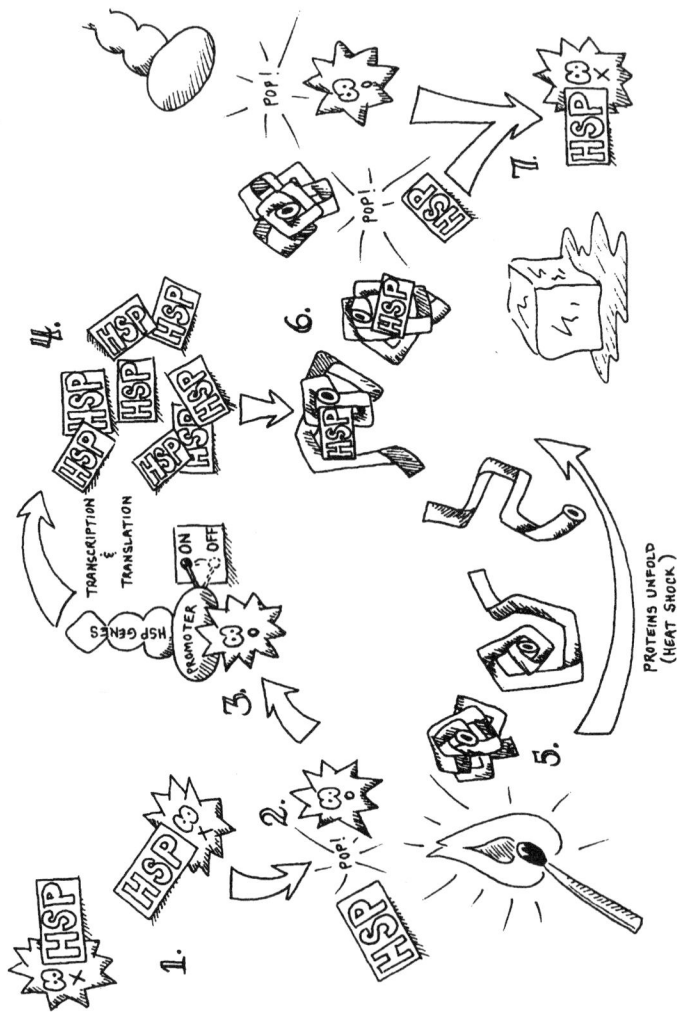

Figure 4.6 Model for the regulation of genes by heat shock proteins (HSP). (1) In the resting state, the protein Shock Factors (shown as star-shapes) bind as single molecules to the small amount of HSP (shown as rectangles) that is present in the absence of stress (the Shock Factors have "X-mouths" here to indicate that they are unable to activate transcription of HSP genes at this point). (2) After heat shock, the Shock Factors (shown as star-shapes with "O-mouths") separate from the HSP (3) and bind to the promoter of the HSP genes to activate their transcription, which eventually results (4) in the production of more HSP. (5) During heat shock, most cellular proteins begin to unfold. The "old" HSP, which dissociated from the Shock Factor, and the "new" HSP (the production of which is induced by the Shock Factor) both bind and stabilize these cellular proteins (6), preventing them from unfolding. Finally, (7) the temperature returns to normal, and the HSP's revert to the single molecule state, no longer bind to the HSP gene promoters, and instead begin to bind up the very Shock Factors that induced their production in the first place. Therefore, HSP gene transcription stops, and the system stabilizes.

level of The Living Staircase. And I'm sure we've only discussed the tip of the iceberg."

"Speaking of The Staircase, Jeff, let's take a quick look back at the types of regulation we've seen in our run alongside FloJo and her fellow competitors. In the mere space of 10 seconds, there is regulation at the cell and tissue levels (muscle and energy metabolism), the protein and gene levels (energy metabolism and heat shock), and the organ and organ system levels (body cooling and anabolic steroids)."

The Finish and a Last Gasp

"To wrap things up, Pearl, we'll replay the last few seconds of the race. FloJo finishes well ahead of the field. She raises her arms exultantly—an unprecedented effort."

"Jeff, we're not done yet. As FloJo slows down to catch her breath and allow her regulatory mechanisms to readjust to the resting state, I'm reminded of breathing regulation. This is a great example of regulation that ties together many of the concepts we've discussed, as well as many of the steps of The Staircase. Do we have time to briefly review it?"

"Shoot."

"The breathing center control tower is at the base of our brains. In much the same way that FloJo's brain translated the signal of the starter's gun and the brain and the rest of her central nervous system then signaled her muscles to move, the breathing center sends a signal to our rib muscles to contract, and we inhale. Now, Jeff, how do you think we keep from continuing to inhale and over-extending our lungs?"

"There must be some sort of feedback involved, but I can't think how."

"You're on the right track. When our lungs are full, stretch sensors feed back nerve signals that inhibit the breathing center. Remember how hydrogen ions, a byproduct of glycolysis, feed back to negatively regulate PFK? Well, pH, a measure of the level of hydrogen ions, remember, also regulates the breathing center."

"What does pH have to do with breathing, Pearl?"

"Well, let's take it step by step, Jeff? First of all: why do we breathe?"

"Uh, . . . to inhale oxygen and to exhale carbon dioxide, right? We need the oxygen to produce energy. Like we discussed earlier, cellular respiration requires oxygen to produce energy, and I think carbon dioxide, the stuff we need to get rid of, is a byproduct of cellular respiration."[10]

---

10. See Chapter 3 on Krebs Cycle and energy metabolism.

"Good, so keeping that in mind, when the sprinters are breathing heavily flying down the track and then cooling down, or for that matter, when you're breathing right now, what should the breathing center monitor in your blood in order to 'tell' you to inhale or exhale?"

"The levels of CO_2 and O_2, I guess. If CO_2 levels are too high, we better exhale and if O_2 levels are too low, we need to inhale."

"Very nice, Jeff. Once again, you've pretty much got it. Actually, we have oxygen sensors in our arteries as a back-up; they only signal the breathing center when we're severely deprived of oxygen. It's CO_2 levels that the breathing center regularly monitors, and here's where pH comes in. When CO_2 enters the blood, it mixes with water to form carbonic acid. As is the case with all acids, it lowers the pH of the solution. In this case, the solution is your blood. The breathing center senses this decrease in pH and interprets it to mean that CO_2 levels are high. So, the breathing center signals appropriate nerves and muscles to increase or decrease the pace and depth of breathing."

"Well, Pearl, like you said, that last example does a beautiful job of tying together the themes of this week's show. It includes feedback regulation involving ions, molecules, organs, organ systems, and ultimately, the organism. I hope those of you in our viewing audience have enjoyed this special edition of *These Sporting Times* with sports physiologist Pearl Garcia as much as I have. I'm sure I'm speaking for our entire audience when I say I'll never be able to watch track and field events the same way again. Thanks very much, Pearl. We'll see everybody again next week on *These Sporting Times*. Good night."

? Q & A with Jeff & Jen

JEN: That discussion of anabolic steroid abuse is sad and scary. It's amazing what some people will do to win.

JEFF: If you think athletes messing up their internal regulation systems is bad, think about how we may be messing up regulation of entire global cycles. Humans are sort of like the anabolic steroids of the environment.

JEN: That's a little much, Jeff. What exactly are you talking about?

JEFF: Remember in the chapter, where Pearl mentioned that one theory of how steroids might artificially enhance muscle mass is that they increase nitrogen stores in our bodies? And, since we use nitrogen to synthesize our amino acids, this might lead to more amino acids, more protein, and more muscle?

JEN: Yeah, but what's your point?

JEFF: Well, we also need nitrogen for making DNA and RNA and other stuff. Where do you think we get that nitrogen from in the first place?

JEN: I don't know. Food, I guess.

JEFF: That's right, sort of. From this chapter on regulation and other things I've read, I gathered nitrogen was pretty important to life. So, I did some research on the subject.

We're part of this global process called the nitrogen cycle. It involves the whole earth, the very top of The Living Staircase. Nitrogen is in the atmosphere, in the water and soil, and in plants and animals. More than one process can add to or subtract from these pools of nitrogen. For example, nitrogen can enter the atmosphere by combustion from factories, by ammonification from animals and the sea, or by denitrification from the soil or sea.

JEN: Okay, so how does all this nitrogen get into our muscles?

JEFF: Well, we can't use straight nitrogen in our bodies. It has to be converted or "fixed" by blue-green algae or bacteria.

JEN: You mean all the nitrogen in us has been through microorganisms first?

JEFF: Yes. The best-known of these microorganisms are the *Rhizobium* bacteria. They live on the roots of legumes like alfalfa and clover in a symbiotic relationship. We eat these plants, or animals that have eaten these plants, to get our nitrogen.

JEN: Interesting. So, how can such a massive cycle involving bacteria, the atmosphere and everything else be regulated?

JEFF: It's impossible to study this regulation in a test tube. But we can look at the regulation of different parts of the cycle. For example, bacterial nitrogen fixation and amino acid synthesis are regulated a lot like glycolysis. At least one enzyme controls each step in the synthesis of different amino acids, and there is feedback control of key regulatory enzymes in the pathway. Also, making amino acids requires ATP. So, amino acid synthesis is tied into glycolysis, since glycolysis makes ATP.

JEN: It seems like, if you worked at it, you could tie the regulation of every cycle on earth into the regulation of every other cycle; which reminds me, you said humans are the anabolic steroids of the nitrogen cycle, and you still haven't told me what you were talking about.

JEFF: Well, let's take the idea that we can look at the entire earth as a living organism. This idea is known as the Gaia Hypothesis. It's only an idea, but the earth-as-organism analogy helps me out here.

Taking anabolic steroids alters the complex web of hypothalamic regulation. How do we do the same with the nitrogen cycle? We artificially add thousands of tons of nitrogen into the cycle. Combustion from factory furnaces dumps fixed nitrogen into the atmosphere, and fertilizers put about half the fixed nitrogen into the soil that all microorganisms together do naturally. When we build on river banks or when we clearcut forests, we cause a large increase in nitrogen compounds moving from the soil into waterways.

JEN: I see what you're getting at.

JEFF: Also, the nitrogen and sulfur emissions from our burning of fossil fuels result in acid rain. Acid rain makes for acid soil. At low pH, nitrogen-fixing bacteria in the soil perform their vital fixing process very poorly, if at all (Just like our muscles can't work at low pH). Remember that these bacteria must fix all the nitrogen we use.

JEN: Sounds like feedback regulation.

JEFF: Exactly. And like you were saying before, all regulation connects at some level. The nitrogen cycle interconnects with the global cycles of sulfur, oxygen, carbon, and phosphorous.[11] The man-made changes in the nitrogen cycle that I was talking about cause an increase of NO_2 in the atmosphere. This could lead to increased ozone depletion. As we keep hearing in the news, the ozone is the layer of atmosphere that protects us from cancer-causing rays of the sun.

JEN: This sounds dangerous. Can the cycle deal with all this extra nitrogen we're making?

JEFF: Good question. The cycle can sort of adapt to our excesses. For example, when nitrogen levels get too high in soil from fertilizing or other factors, fire can burn off the nitrogen, it can be permanently fixed in organic matter, or bacteria can "denitrify" it.

And I guess genetic variants of bacteria may arise that can fix nitrogen in acid soil. But we don't really know how our interference will affect other parts of the cycle, like humans, for example. You could say we're doing the experiment on ourselves right now, as we speak.

---

11. See the discussion of global cycles such as these in Chapter 3.

As with steroids and hypothalamic regulation, we can see the *initial* effects of interfering with the nitrogen cycle, but more subtle effects are less clear.

RELATED RESOURCES

Getup and Go, K. Moore, *Sports Illustrated*, July 25, 1988.

Lore of Running, 3rd edition, T. D. Noakes, Champaign, IL: Leisure Press, 1991.

The History of Synthetic Testosterone, J. M. Hoberman and C. E. Yesalts, *Scientific American*, Feb. 1995, p. 76.

The human heat shock protein hsp70 interacts with HSF, the transcription factor that regulates heat shock gene expression, H. Abravaya, M. P. Meyers, S. P. Murphy, and R. I. Morimoto, *Genes and Development* 6, 1992.

HSP70 and other possible heat shock or oxidative stress proteins are induced in skeletal muscle, heart, and liver during exercise, D. C. Salo, C. M. Donovan, and K. J. A. Davies, *Free Radical Biology and Medicine*, 11, 1991.

Nitrogen in Terrestrial Systems; Questions of Productivity, Vegetational Changes, and Ecosystem Stability, C. O. Tamm, New York: Springer-Verlag, 1991.

HACKERS

Theme

The effects of natural selection can be seen in most levels of The Living Staircase, from molecules to populations. Natural selection is the process that causes populations to evolve. By adapting organisms to different habitats, selection has created the great diversity of life around us.

Examples

■ RNA replication
■ Mutation and protein synthesis
■ Human Immunodeficiency Virus
■ Fat deposition in adipose tissues
■ Heterostyly in primroses
■ Population genetics of the mussel *Mytilus edula*

March 2, 1988, started out as a regular day—I woke up at 7:30 a.m., showered, and sat down with a cup of coffee at my computer by 8:30. Then I hit the "on" switch. The happy mac face appeared, and I expected the familiar desktop display to follow. Instead, my computer produced a picture of earth with this message:

RICHARD BRANDOW, publisher of MacMag, and its entire staff would like to take this opportunity to convey their UNIVERSAL MESSAGE OF PEACE to all Macintosh users around the world.

"Peace," I thought. "That would be nice." Then I clicked the mouse and went about my work. I had no idea that I had participated in an early experiment on the creation of life.

The peace virus, as it came to be called, made the headlines in computing magazines for weeks. The virus was harmless—it self-destructed the following day—but computer users immediately recognized the potential danger of such programs. "Computer terrorism" they called it. I was wary of the danger; but as a biologist, I was also intrigued. The peace virus had shown a characteristic of life. It was capable of reproduction.

Looking into the background of the virus, I learned that it was created by Drew Davidson, a Canadian hacker who was challenged to write a program that would silently spread wherever it could, whenever it could, and self-activate on March 2. Davidson did a technically brilliant job, and presented a floppy-disk copy of the virus to the publisher of MacMag, Richard Brandow. In December of 1987, Brandow left the virus on two hard drives in a computer store in Canada. It spread from there like a biological virus, stealthily jumping from infected to uninfected individuals (hard drives) during opportunistic contact (modem connections and floppy-disk contact). Three months later, 350,000 of us got the message. Computer viruses reproduce and spread at an impressive rate.

However, to say that computer viruses reproduce is not to say that they are alive. There are several additional characteristics of life: growth and development, energy utilization, homeostasis, responsiveness to environmental cues, and evolutionary adaptation. Computer viruses meet some of the criteria (reproduction, energy utilization, and responsiveness to the environment), but fail in others (growth and development, homeostasis).[1] What about the criterion of evolutionary adaptation? Are com-

---

1. Interestingly, biological viruses also fail in some of the criteria. Because they cannot reproduce on their own (they have to infect a cell) and go through long phases in which energy input is not necessary, they are not considered living entities by most biologists.

puter viruses capable of this? In this chapter, we will examine the question by focusing on adaptation in natural populations, and the process that produces it—natural selection. What conditions are necessary for selection to occur? How strong is its effect on The Living Staircase? Could selection affect computer viruses? To begin, let's cover the basics.

Natural Selection

Selection occurs when a population of organisms contains variation, and some variants are more successful at reproducing than others. As a result, the population changes over time. As a hypothetical (but perhaps realistic) example, imagine a dorm that is colonized by roaches (Figure 5.1). During the summer, they live an easy life—the roaches grow and reproduce without having to evade the shoes of students. Short-legged (slow) or long-legged (fast), they all have the freedom to live and breed, and they are equally successful at it. The return of students in late August changes things, however, and a nine-month period of selection begins.

Short-legged roaches scramble slowly, get hammered more often, and thus reproduce less. As the school year progresses, the population grows (because, you know, you can never beat 'em) and also changes—long-legged roaches have higher survival and reproduce more, and thus make up a greater and greater proportion of the overall population. Selection, the process that created the intricate structure of The Living Staircase, has taken place.

Biologists often describe selection in terms that imply a conscious process—long legs are "favored," short legs are "selected against." These are forms of speech that are used for convenience, not for an accurate depiction of how selection works. Selection can also be stated as part of a formal syllogism, without implying any deliberate activity:

IF

a. the members of a population have different traits, and

b. some traits lead to higher reproduction than others, and

c. the offspring inherit the traits of their parents,

THEN

d. over time, some traits will become more common in the population while other traits become rare.

Figure 5.1 Selection occurs everywhere in nature, including dormitories. In this dorm, the roach population contains long-legged and short-legged varieties. Each roach that makes it to adulthood gives rise to two offspring. However, short-legged roaches are more likely to be chased down and stomped by students. As the school year progresses, selection against the short-legged causes the population to change.

Conditions *a, b,* and *c* define what is necessary for selection to occur; *d* is the result of selection—evolution. In other words, evolution is an emergent property of selection. And as explained in Chapter 1, an emergent property can be studied from different perspectives. One approach is to focus on selection in action; to do this, a biologist would hang out in a dorm and record how often a student stomps a roach, and what type of roach it is. In most cases, however, the biologist only sees the outcome of selection, the so-called "Ghost of Selection Past." Visiting the dorm at the end of the school year, the biologist would observe that the population has very few short-legged roaches and devise an explanation based on past selection. In this chapter, we will use examples from both perspectives.

Biological populations usually satisfy conditions *a, b,* and *c*, so they are typically under some form of selection. But do computer viruses meet the criteria? Condition *c*, certainly—when the peace virus replicated, all of its "offspring" contained the same message. Condition *a*, however, was not met. All copies of the peace virus were identical, and this is generally true of computer viruses. We will return to condition *a* later, and learn why it is essential for evolution to occur.

What about condition *b*? Upon first learning about computer viruses, I questioned whether they had the second characteristic—are some better at reproducing than others? If you are comparing viruses that arise from different sources (not the descendants of a single virus) the answer is a definite "yes." This was proven by the "Core Wars," an innovative game played in the early 1980s by young computer scientists (hackers with degrees) at AT&T. After the offices closed, the players wrote self-replicating programs—viruses—and then released them into an isolated computer where they competed for memory space. On any night, one virus would rise to championship status—defined as having commandeered the greatest amount of space—and, no doubt, its creator enjoyed a free beer on the way home. The Core Wars showed that computer viruses differ in their capacity to reproduce.

As the Core Wars continued, the programmers created versions that went a step beyond taking up space—they destroyed other files to make extra space for themselves. Eventually, one of these escaped the confines of the game computer and made its way into the AT&T mainframe computer. The company was shut down for the better part of a day, and following that, the Core Wars were discontinued. (Coincidentally, some young computer scientists were given pink slips at the same time.) Today, the Wars can be found in a different forum—one that seeks to unravel the earliest steps in the evolution of life. This form of the "game," which is now a serious scientific experiment, is played by biochemists that specialize in the study of **RNA**. A researcher puts several different strands of RNA into a test tube and furnishes all of the other required materials for replication (the molecular equivalent of reproduction), and lets the process take off. Over a couple of hours, the amount of RNA in the vial rises as replication takes place. In addition, one type of RNA comes to predominate. This is the RNA that was best suited to the environmental conditions, like pH and temperature, in that experiment. It was "favored" by selection, while the other RNA variants were "selected against."

The RNA experiments simulate an event that took place about four billion years ago—the emergence of natural selection on our planet. Why RNA? Because it was probably the first reproductive structure on the planet, and reproduction is a key requirement for evolution to occur. Entities (whether biological or computerized) that do not reproduce cannot respond to selection.

Selection and Diversity

Both the Core Wars and the RNA experiments demonstrate selection in its purest form. A population of self-replicating entities is released into a "habitat" with ample space, and one variant rises to the top by outreproducing the others. Obviously, selection does not operate this way in nature; if it did, there would be little biological diversity—we would see a single "best type" of cockroach, not the 3500 species that have been discovered so far. In nature, selection actually produces diversity in the process of adapting organisms to various habitats. Different traits are favored in different locations. A famous example of this comes from H. B. D. Kettlewell's study of the peppered moth in England.

In the early 1950s, Kettlewell became interested in *Biston betularia*, a species of moth that comes in two colors—plain dark brown (the melanistic phenotype) and white with dark speckles (the peppered phenotype). Kettlewell noticed that 98% of the moths in his town of Manchester were melanistic, while in the surrounding countryside, the peppered phenotype was more common. In other words, the moths were distributed in patches—Manchester was a primarily dark patch and the outlying areas were peppered patches. He knew that he was observing the Ghost of Selection Past. But why had selection produced this distribution?

Kettlewell hypothesized that the patchy distribution was related to pollution. During the day, *B. betularia* rests on tree trunks, and relies on camouflage to protect itself from birds. Manchester moths had to blend into pollution-covered bark, so selection had favored the melanistic phenotype in the city. In the countryside, peppered moths were favored because they blended into lichen-covered bark.

To test his hypothesis, Kettlewell released melanistic and peppered moths in a soot-blackened woodland. The peppered moths were victims of birds much more often than the melanistic ones. He also examined museum collections that dated back a century, and found that Manchester had been a "peppered patch" until 1900, and then in a mere 50 years, selection had restructured the population to favor the melanistic morph. This timing coincides with the industrial revolution, which polluted the Manchester woodlands.

Kettlewell's study demonstrated that natural selection can lead to diversity in nature. Selection in the operating environment of a computer, however, is quite different. During the Core Wars, a single virus rose to the top because its habitat, a computer hard drive, was homogeneous—every section of it looked like any other. Natural habitats, however, are heterogeneous, so a phenotype that is favored in one area may be selected against in another. As organisms adapt to different habitat patches, variety increases. There are hundreds of stories like that of the peppered moth, and taken together, they have convinced biologists that selection has produced the incredible diversity of nature.[2]

Levels of Selection

Selection is a challenging concept to understand because it **operates** at the level of the individual, but produces **effects** at other levels. Individual moths in Kettlewell's study lived or died, reproduced or did not; but it was the population that changed. Does selection affect the other levels of The Living Staircase? Certainly. As selection occurs, survivors contribute their genes to the next generation. Because genes affect all aspects of an organism, from its molecules to its behavior, the effect of selection can be detected at many levels—even in the structures of DNA and RNA.

Selection at the Molecular Level

DNA and RNA interact to specify which proteins a cell is making at any particular time. Before getting to the subject of selection, we need a brief overview of these interactions. When a cell needs protein X, the region within a DNA molecule that carries the genetic code for protein X is copied; the copy, messenger RNA (mRNA), leaves the nucleus to take the instructions to the cytoplasm (Figure 5.2). mRNA carries the information by virtue of its structure; the molecule is made of subunits abbreviated A, C, U, and G, and the sequence of these determines the message delivered.[3] An mRNA molecule that reads UUAGCGCCGUAC carries a different message than one with the sequence UUUGGCCAUUAA.

Once in the cytoplasm, the mRNA is interpreted by a ribosome, which attaches to mRNA and begins to "read" the message three nucleotides at a time. Each triplet (a "codon") of RNA letters is the code name for an amino acid. For example, a ribosome would read the first sequence in the previous paragraph as UUA GCG CCG UAC—

---

2. The concept of diversity, and many examples, are given by the famous classifier, Carolus Linnaeus, in Chapter 7.
3. See Chapter 6 for an in-depth description of the DNA-RNA interaction.

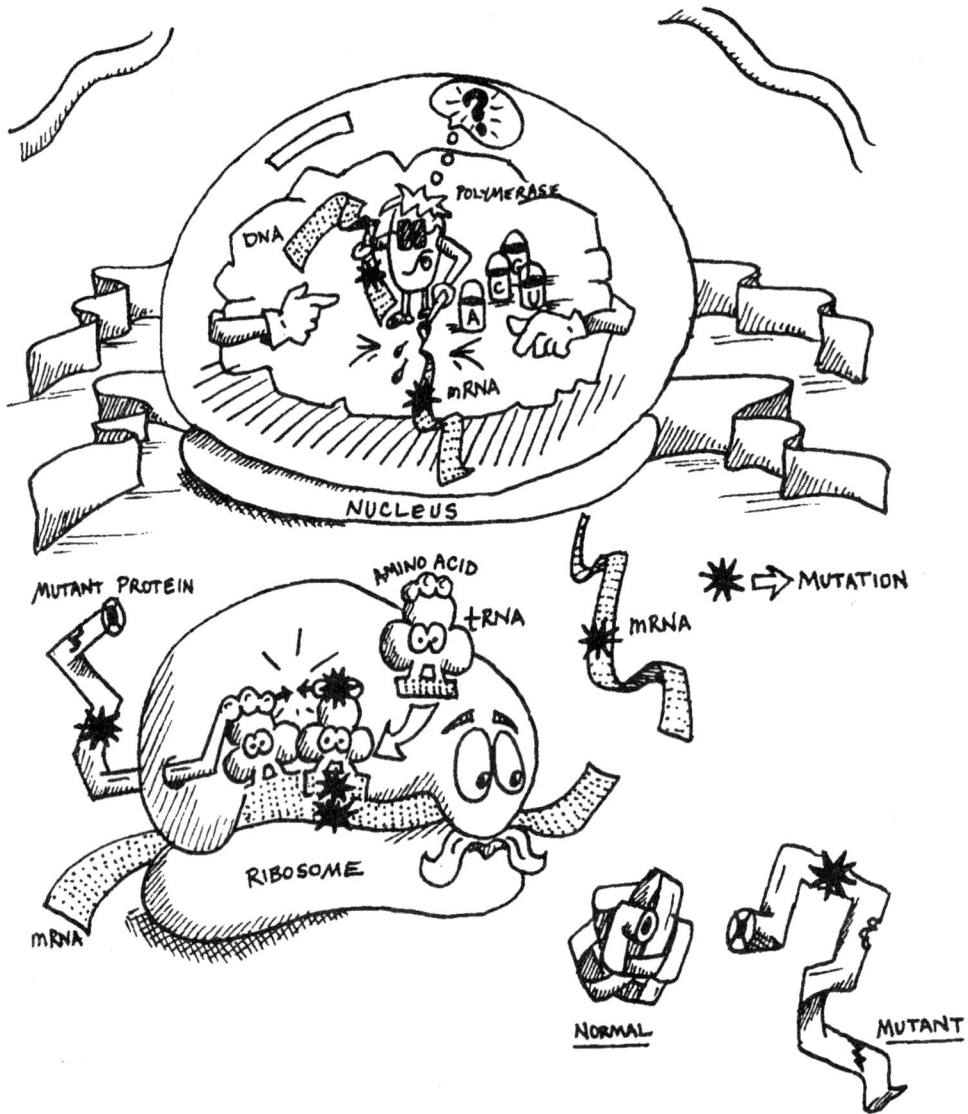

Figure 5.2 How mutations affect protein synthesis. In a eukaryotic cell, the DNA molecules are in the nucleus. To send information into the cytoplasm, a section of a DNA molecule (i.e., a gene) is copied into RNA language. The RNA molecule leaves the nucleus, where it is read by a ribosome to make a protein. A mutation in the DNA molecule is copied into the RNA message, and can lead to the production of a mutant protein. The process is similar in a prokaryotic cell, except that there is no nucleus for the RNA molecule to leave.

these four codons spell out the identities of four amino acids that should be strung together to make the protein. (This case is not realistic because most proteins are hundreds of amino acids long, not just four.) If you have followed me up to this point, we can now move on and see how Ghost of Selection Past is present in DNA and RNA.

When the genetic code was being deciphered in the 1960s, molecular biologists discovered that some codons are synonyms, meaning that they specify the same amino acid. The codons UCU, UCC, UCA, and UCG all mean the same thing to a ribosome: "attach a molecule of serine here." If you look over these synonymous codons, you can see that they differ only in the third letter. Apparently, this position in a codon is less important than the first two. This means that a mutation (a substituted letter) in the third position of a codon will not result in a change in the instructions to the ribosome, but a mutation in the first two letters will.

Now, then. Figure 5.3 shows a consistent trend found by molecular biologists. If you compare the gene for a particular protein in a thousand humans, you will find mutations in the third position of codons much more often than in the first or second positions.[4]

Selection provides a robust explanation for this finding. Mustations in positions one or two lead to nonfunctional proteins, and past humans that had such mutations had lower survival; thus, the mutations were not inherited, and are not present today. Mutations in position three tend to be neutral, so they accumulate over time.

Selection and Mutation

This example of selection might leave the impression that all mutations in positions one or two are bad. While it is true that the majority of such mutations are selected against, there is always the possibility that favorable or neutral mutations could arise in these positions. In fact, a low level of mutation is essential for evolution, because it is mutation that provides the initial variation that selection works upon— condition *a* in the three criteria for selection. Consider the moths in Manchester prior to the Industrial Revolution, when the peppered phenotype was abundant. If the population had been composed entirely of the peppered moths, and the gene for melanistic coloration was entirely absent, the moth population could not have evolved to become darker later on. Mutation is essential for evolution because it creates a variety of traits for selection to work upon.

For computer viruses, mutation requires the assistance of hackers, and unfortunately, most of them are driven by malicious intent. One virus in particular, the nVir virus, seems to take pleasure in hunting me down after every evolutionary advance.

---

4. Put another way, letters one and two of a codon are "saved" or kept the same during evolution, while letter three is not. Similarity of structure has been a consistent trend throughout evolution, as Chapter 6 details.

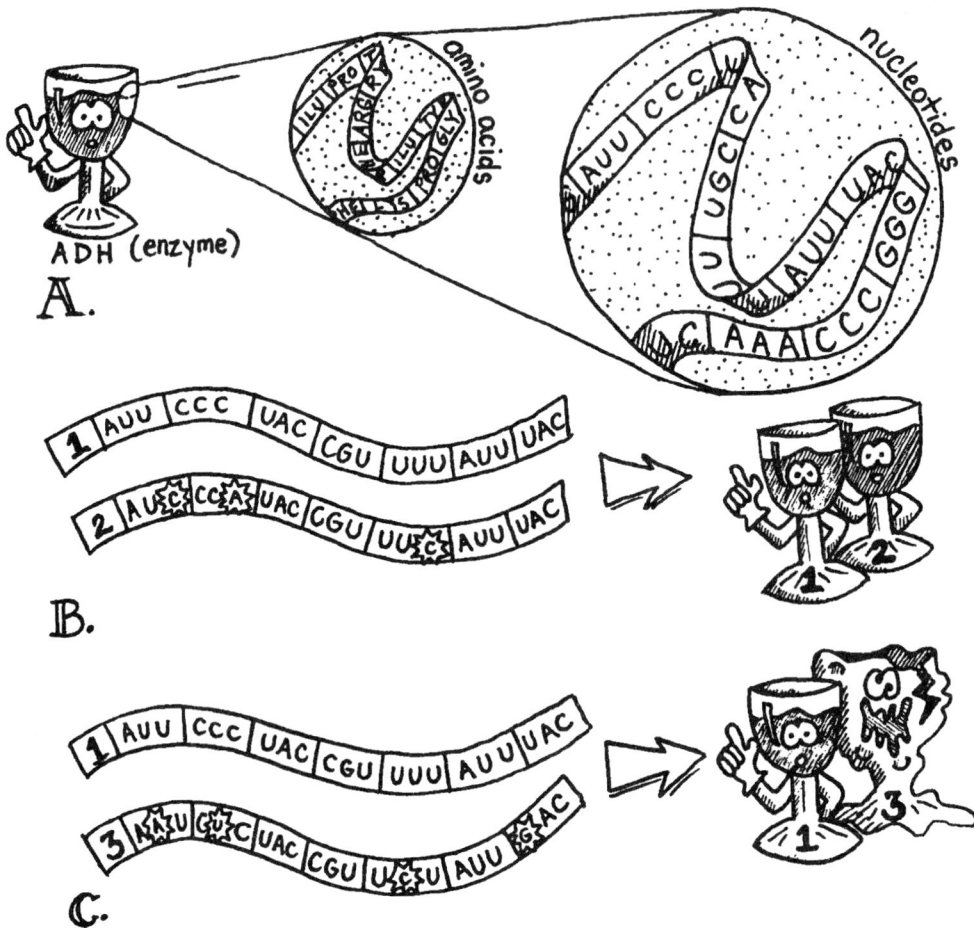

Figure 5.3 Alcohol dehydrogenase (ADH) is an enzyme that can be found in the cells of virtually all organisms. (A) The chain of amino acids that make up ADH must be precise for the enzyme to function. (B) The top line shows the human RNA code for ADH; in the bottom line, some codons have been altered by mutation. These alterations would not affect the structure of ADH, however, because the changes are only in the third letter of each codon. When comparing individuals within a population, third-letter mutations are fairly common. (C) Changes in the first or second letter of a codon result in a nonfunctional form of ADH. These are selected against except in the rare case where the molecule's function is improved.

The first version caused infected programs to beep when they began running. It was annoying. The second version, modified by someone with a devilish sense of humor, beeped and then displayed the message "Don't panic." It was vexing. And the third version gave no beeps or messages, but randomly selected a system file and destroyed it. It was crippling!

nVir is evolving, but only with the help of hackers who make the changes that lead to more successful varieties. Because the virus does not mutate spontaneously—on its own—there is no variation within a population (condition *a* for selection), and its evolution is slow. Once an infection is discovered, it is easy to eliminate with a virus-protection program. In contrast, biological viruses undergo spontaneous mutation at a high rate, evolve quickly, and stay ahead of our efforts to control them. HIV, the virus that causes AIDS, is a case in point. Research on it has revealed selection in action at the cellular level.

Selection at the Cellular Level

HIV is extraordinarily virulent. Those who are infected eventually die from the association with it. But "eventually" is the key word here. In most victims, a period of about ten years passes between first contact with the virus and death. During that time, a process of selection is operating within the person's immune system.

HIV specifically seeks out and infects a group of cells called Helper T- (T_H) cells. T_H cells are a central part of every person's immune system—they coordinate the immune response, which does the dirty work of killing invaders. They can be thought of as the command posts from which a battle is directed. When HIV eliminates the T_H cells, the immune system is left uncoordinated, leaving the person much more vulnerable to diseases.

To see how selection affects the progression of AIDS, let's track the life of a single HIV particle (a "virion"). When the virion first infects a person, it is targeted for elimination by the immune system. However, the immune response is rather slow, taking several days to produce antibodies that can kill the HIV invader. During this time, the virion moves through the circulatory system until it finds a T_H cell. T_H cells have no defense against the virus—HIV easily attaches and injects its core. Interestingly, the outer shell of the HIV virion remains outside the cell, and is later attacked and destroyed by antibodies. But the antibodies are too late—by then, the genetic code for HIV is hidden away within the T_H cell, and is in a perfect refuge. It cannot be detected or destroyed.

Once in safe surroundings, HIV replicates at an impressive rate, doubling its population every two days. At first, this is not a problem—there are many uninfected T_H cells to sustain the immune system's function. As the viral population grows, however, it infects and kills an increasing proportion of the T_H cells (Figure 5.4). One

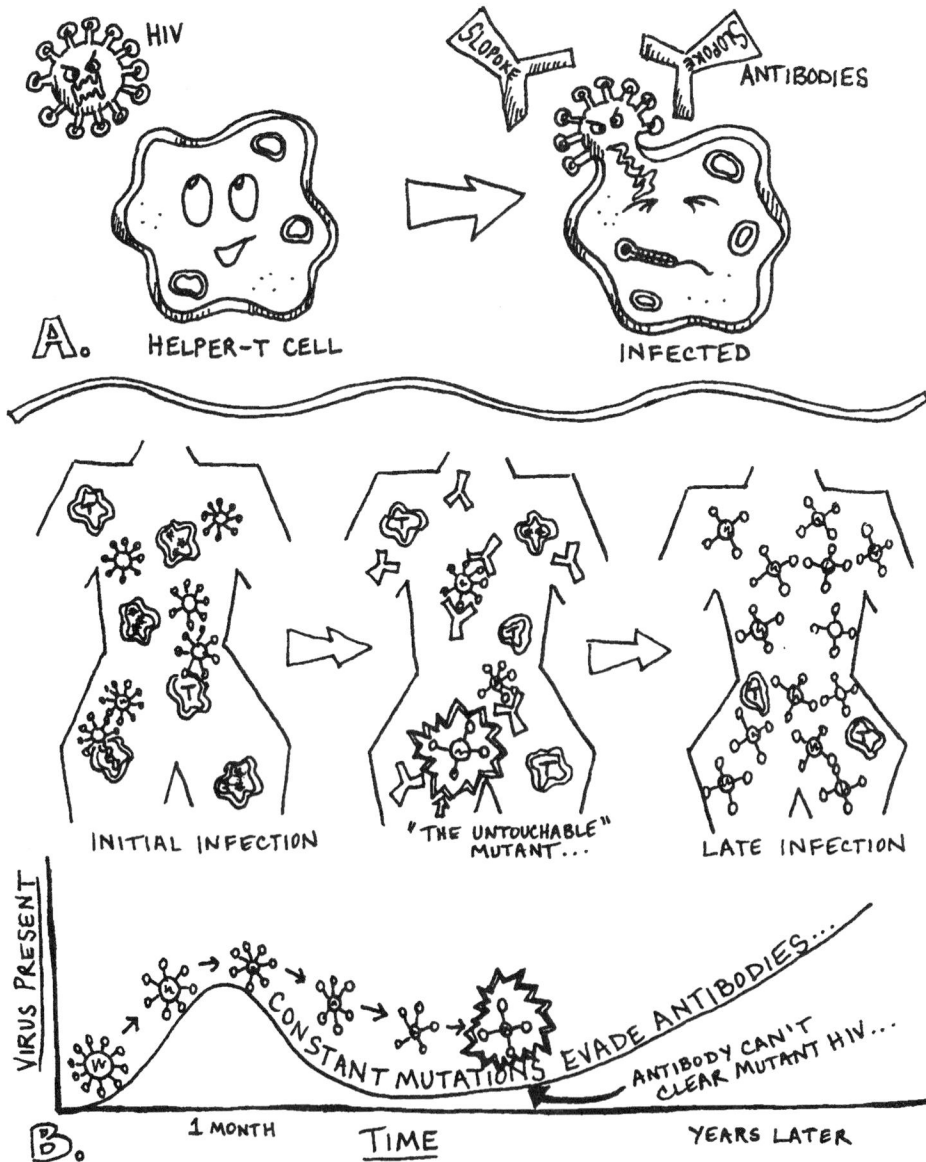

Figure 5.4   Selection occurring in the immune system of an HIV-positive person. (A) HIV attaches to Helper-T cells and injects its RNA. During the initial infection, the person's immune response is too slow to prevent infection. After infecting the cell, the virus replicates itself. (B) Its rate of mutation is so high that each new virion is different. Most will be attacked by the antibodies that were produced after the initial infection, but eventually, one is produced that is not recognized by the antibodies. This virion infects, and ultimately kills, other T$_H$ cells.

month after the initial infection, the virus reaches an incredible killing rate—it is estimated to eliminate two billion T_H cells a day, and the victim is barely able to replace the lost cells. It is at this point that selection comes in.

When HIV copies itself, most of the copies are attacked and destroyed by antibodies that were produced in response to the original infection. But the virus has a mechanism to outmaneuver its enemies—it mutates at a very fast rate. So fast, in fact, that it is believed that every new virion leaving an infected cell is unique in some small way. Eventually, some virions arise that are different enough to bypass the existing antibodies. These have a few days to spread before the immune system can design a new antibody to attack them. As this scenario continues for years at a time, the virus becomes increasingly virulent, and eventually reaches a point where it kills T_H cells faster than the body can replace them. The victim is left with a weak ability to fight off disease, and is killed by an illness that most of us would find trivial.

As humans, we have an immune system that is astonishing in its ability to recognize and destroy the diverse pathogens in our environment. How ironic, then, that the immune system in an HIV-positive person ultimately brings itself down—for it is the immune system that does the selecting here. By removing less virulent forms of the virus, it ultimately selects in favor of the virions that can outwit it.

In the early 1980s, there was a computer virus that bore some resemblance to HIV. The ASD (acquired system deficiency) virus spread cleverly, hidden in a "Trojan horse" program. The horse, known as Sex.exe, was posted on several bulletin boards around the country. Users downloaded Sex.exe, and while running it, were entertained by a depiction of humans engaged in interesting activities; but in the invisible background, the ASD virus sought out and destroyed a critical system file. Later, the infected computer would inexplicably crash. Victims of ASD learned a lesson about infection control—you shouldn't toy with software that is not shrink-wrapped. (There is an important parallel with HIV, if you think about it.)

By virtue of the process of selection in an HIV-positive person, researchers can find several different strains of the virus in a person that was originally infected with one strain. Once again, this is an example of selection leading to an increase in diversity in a population—only this time, it is a population of virions within a human being. Selection has also produced the diversity that we see among ourselves; in fact, one of our secondary sexual characteristics is attributable to the Ghost of Selection Past.

Selection at the Tissue Level

Just under the skin, there is a layer of cells called adipose tissue where we store our excesses—the intake of calories that exceeds the body's needs. At birth, our adipose cells are skinny and star-shaped; but they begin to expand as globules of fat are deposited inside. By the end of infancy, we all have the baby fat that is adored by those around us, and to which we are happily oblivious. Thereafter, we follow a genetically determined path that leads to distinctly different patterns for men and women. The tissue is thicker in different places, leading to different shapes. When men gain weight, they deposit most of the fat in the adipose around their midsections, as any visitor to a country bar can testify. Women deposit fat further down, around the hips and thighs. Also, men and women differ in the **amount** of fat deposited: pound for pound of body weight, women have more. This is true for all races, in all regions of the earth.

Why would it be so? Since this is the result of past selection, we have to guess at a reason—but a good one comes to mind quickly. Women have a periodic metabolic expense that no man can match. It is pregnancy. And following that, milk production for nursing is another drain on energy reserves. To deal with this, our ancestral mothers evolved to stockpile more fat than their male counterparts. Apparently, the Ghost of Selection Past was harder on skinny women than on skinny men.

This being built into the genes, women are stuck with the tendency to put on fat more easily than men, and to keep a greater fat reserve. Is it a curse? Many would answer in the affirmative, but there is a payoff—the thicker adipose layer gives women higher survival under starvation conditions. History gives ample proof of this, for there has been no shortage of human crises. But one of my favorite stories is that of a wagon train in the Old West.

In the 1840s, thousands of American settlers hurried across the western U.S. to settle in California, the territory that had just been won from Mexico. The Donner Party was a small wagon train that branched off from the main path in Utah, hoping to cut off a month's worth of travel time. But as usual, the short cut ended up being longer, and the members of the Donner Party were the last immigrants of 1843 to attempt to cross the final challenge—the Sierra Nevada mountains.

Shortly after they started the ascent on October 28, it began snowing. The snow reached a depth that made travel impossible when the lead wagon was a mere five miles from the summit—only 25 miles to the west, green pastures and complete safety awaited; but the Donner Party had to retreat and encamp in the sheltered coves around a nearby lake. There, they waited and hoped for a break from the harsh winter. The break never came.

Food ran out within a month. After that, the people boiled and chewed the skins of their butchered livestock. When starvation seemed imminent, one group decided to take their chances on hiking out. They estimated that the hike of 25 miles could be accomplished in seven days if the weather was cooperative. So on December 16, a group of five women and ten men set out, each carrying seven days' rations.

Repeated blizzards disoriented the party, the weaker members of the group straggled behind, and the trip took a full 35 days. The testimonials of the survivors indicate that the women were in better shape than the men; a typical day's hike saw the scout, a man, in the lead, the women second, and the rest of the men trailing. Several of the men died and were consumed by the others (except for two Native American scouts who refused to eat human flesh). By the time the hikers reached the first ranch on the western slope, eight of the ten men had died. All of the women survived.

The story of the Donner party is not only a historical record of selection in action, but also a testimonial to the frontier spirit. Their path over the mountains is now crisscrossed by Interstate 80, and the site where they spent the winter of 1843 is a historic monument—complete with a picnic area! To me, it is a memorial to beaurocratic oversight.

Selection at the Organ Level

So far we have reviewed examples of selection at the level of molecules (DNA and RNA), cells (T_H cells), and tissues (the adipose layer). As we work along The Living Staircase, our next step should be an example at the organ level, and we will turn to reproduction in plants to see it.

Flowers contain the reproductive organs of plants, and surprisingly, the organs can be clearly classified as male or female (Figure 5.5). The male organs, stamens, contain the pollen that is analogous to sperm. The female organ, the carpel, is in the center of the flower, surrounded by stamens. The tip of the carpel has a sticky surface to receive pollen. When a pollen grain is in position on a stigma, it germinates and digests a tunnel down the style to the ovary, where at least one egg is waiting.

In some plants, the carpels and stamens are markedly different in length.[5] Primroses are a good example—within the species, there are two types of flowers. One type has short stamens and a long style, and the other type has the opposite arrangement. This seems to be another case of selection producing diversity. In this case, however, the flower types do not occupy different habitat patches; they are often found right next to each other. So why are there two types?

Charles Darwin investigated this question by cross-pollinating the flowers. When he took pollen from a long stamen and placed it on a long stigma, the mating was successful—he got viable seeds. Likewise, a short-to-short mating worked. When he tried the other crosses (long-to short or short-to-long), however, the matings were only half as successful, as judged by the number of seeds that developed. Darwin deduced that there are two mating types in primroses, and that selection had modified

---

5. The vast amount of diversity in plant reproductive systems is described in Chapter 7.

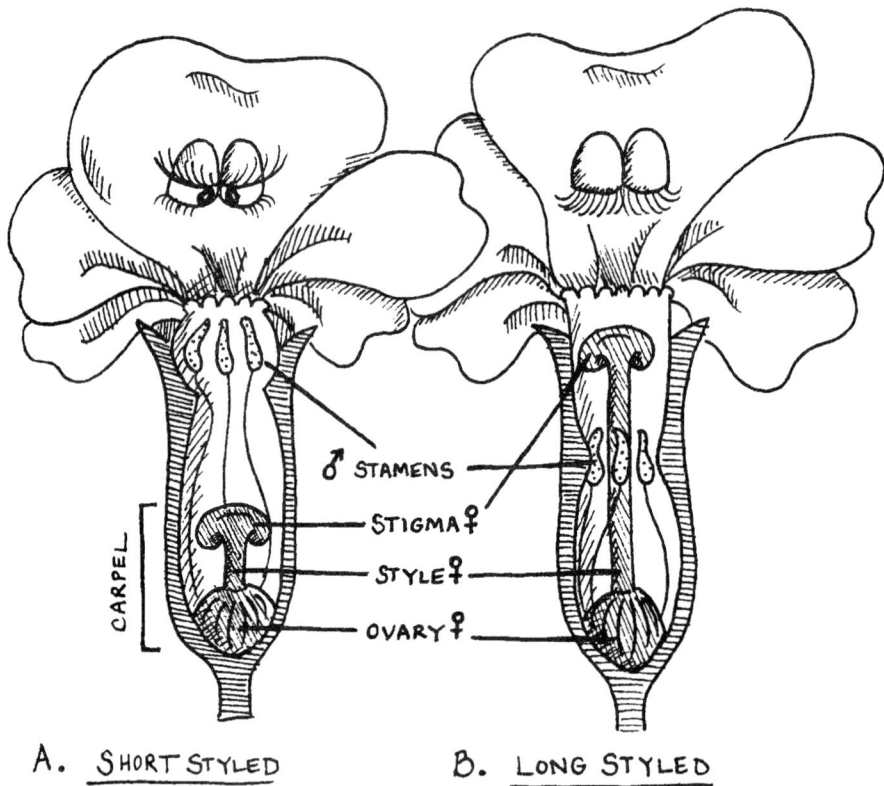

Figure 5.5 Many flowers are bisexual, meaning that they contain male and female reproductive organs. The stamens contain pollen, which is the botanical equivalent of sperm. (Actually, a nucleus in the pollen grain becomes two sperm cells just prior to fertilization.) The carpel is the female organ; it has a sticky surface where the pollen lands (the stigma), a chamber housing the eggs (the ovary), and a stalk connecting the two (the style). Primroses have two types of flowers. (A) One type has long stamens and a short style, and (B) the other type has short stamens and a long style.

the flower structure to reduce the chance that pollen would be transferred from a short stamen to a long stigma, or vice versa.

This is a nice explanation, provided that it applies when pollen is transferred by insects rather than by an elderly biologist. Observations have shown that the long/short dichotomy is an adaptation to take advantage of the different insects that come in search of nectar. When an insect with a long proboscis, like a butterfly, visits the flower, it stands on the petals and snakes the proboscis deep inside, where the nectar is located. Because it is standing out on the petals, the butterfly will only pick up pollen from a flower with long stamens. And when it flies to a primrose with long stigmas, the appropriate transfer of pollen is accomplished.

In contrast, insects like bumblebees scramble right past the long reproductive organs, and thus accomplish the short-to-short pollinations. Flowers with stamens or stigmas of intermediate length would waste their reproductive potential every time their pollen was transferred to an incompatible flower.[6] By favoring two lengths of reproductive organs, selection has provided a simple solution to the complex problem of pairing the appropriate primrose flowers.

Selection at Multiple Levels

In exploring examples of different levels of The Living Staircase, we might have committed the fallacy that this entire book attempts to destroy—that is, giving the impression that a process in biology affects cells, organs, populations, etc., separately. As a final example of selection, then, I'd like to show how selection operating at the organismal level can produce ripples throughout The Living Staircase. The example involves a mussel, *Mytilus edula*, which lives a life that is unfamiliar to most of us. However, the species is not a stranger—it is the guest of honor, so to speak, in mussels marinara.

Mytilus edula is a prolific species; it occupies cool coastal waters throughout the northern hemisphere. The life of an adult is rather lonely; having attached itself to a rock as a juvenile, it stays forever in one place. This makes mating somewhat of a challenge. Direct contact between the sexes is unlikely; so to reproduce, males and females release their gametes into the ocean currents.[7] A female produces an astonishing 25 million eggs per reproductive season, and males release an even greater number of sperm. Floating in marine currents, the gametes unite to become zygotes. As the zygotes are developing into larvae, they are swept over long distances. Every spring, millions of larvae end up in bays along the continental shores of North America. Those that land in Long Island Sound have received considerable attention from evolutionary biologists, in addition to hungry New Yorkers.

The mussels that settle into Long Island Sound fall into three categories—call them A, B, and C—that are evenly distributed throughout the area (Figure 5.6). In other words, the population is homogeneously distributed; mussel type A is just as likely to be found in any particular spot as is B or C. The larvae attach and begin maturing into adults at the same time that selection begins to operate on them.

The three types of *Mytilus* differ in the "lap" gene. *Lap* A, *lap* B, and *lap* C code for the enzyme aminopeptidase, which breaks down proteins into amino acids. The

---

6. Other examples of intricate adaptations for pollination are given in Chapters 7 and 8.
7. Gametes released in this way use chemical messages to find each other. Chapter 9 describes this amazing form of communication.

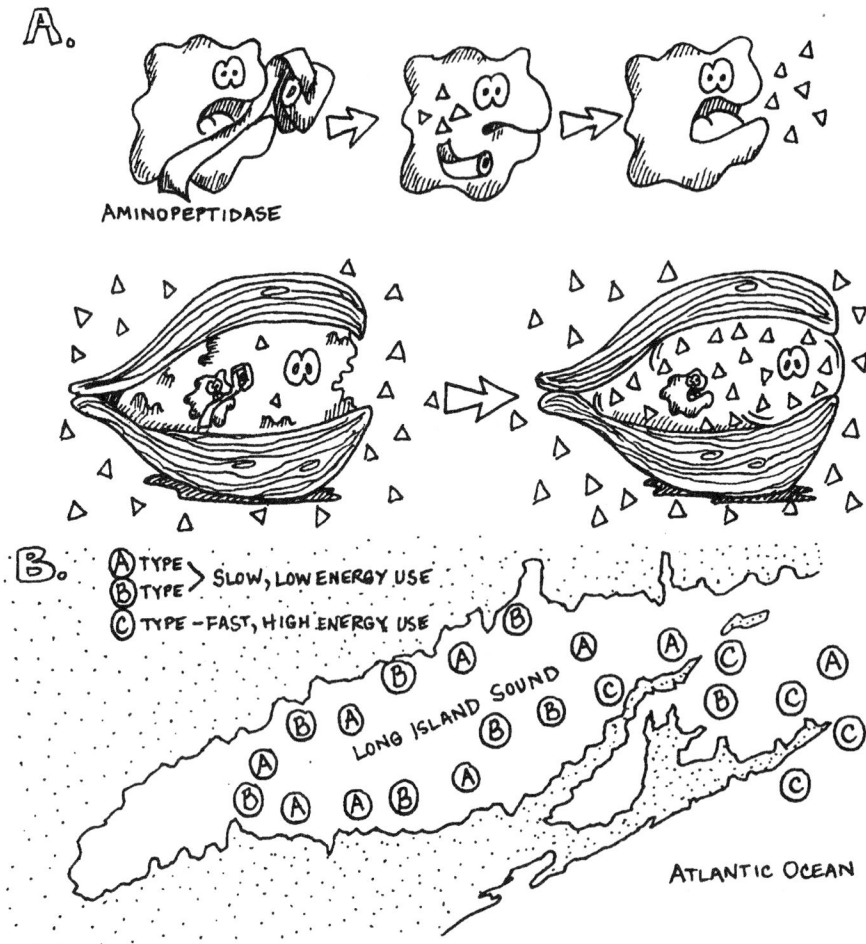

Figure 5.6 (A) The role of aminopeptidase in a mussel. When the salinity of the water around a mussel increases, aminopeptidase breaks down proteins inside its own body. This produces amino acids that match the solute concentration inside the body with that of the surrounding water. (B) There are three types of aminopeptidase in mussels. In Long Island Sound, aminopeptidase type C is most common in mussels that live near the mouth of the bay. Mussels with type C are not well adapted to live in the inner reaches of the Sound, and thus are less common in those areas.

three types of aminopeptidase are slightly different in structure, and carry out the breakdown of protein with different efficiencies, i.e., with different ATP costs.[8]

Aminopeptidase enables a mussel to match the concentration of its body fluids to the concentration of the seawater it is in. This prevents the organism from shriveling like a raisin. Think it over for a moment—suppose that a mussel in Long Island Sound is enjoying a fine day in moderately salty water, when a tide comes in from the ocean. The water around the animal becomes saltier, and by osmosis,[9] the mussel begins to dehydrate. In such a case, aminopeptidase breaks proteins down into amino acids so that the solute concentration inside the organism is equal to that of the water around it. Thus, osmosis will not dehydrate the organism.

Aminopeptidase type C is very efficient; compared to the other two types, it accomplishes the same goal while requiring only 80% as much ATP. In the middle of spring, the three types of *Mytilus* are distributed randomly in the Sound; but by the end of fall, selection creates a different distribution—from the ocean side of the Sound inland, the type C mussel becomes rarer and rarer. This is selection in action—the type C mussel is selected against, and types A and B are favored.

This pattern of distribution indicates that salinity may be a major factor in determining whether a type C mussel lives or dies. The most saline water is at the mouth of the bay, and the least saline water is at the point where the East River empties into the Sound. *Mytilus* type C, the mussel with the most efficient form of aminopeptidase, does best on the ocean side of the Sound, and worst on the river side. Why?

On the ocean side, the changes in salinity brought on by the tides are abrupt, and aminopeptidase type C is faster in adjusting the mussel's internal solute concentration. The other two types of mussels do not handle the rapid shifts in salinity as well, and suffer higher mortality.

On the other end of the Sound, mussels with aminopeptidase type C are selected against. Apparently, they use ATP too quickly. Aminopeptidase C works faster, and thus uses up a mussel's ATP stock rapidly. For most of the year, this isn't a problem; but in the fall, when mussels throughout the Sound are under starvation conditions, the rapid depletion of ATP pushes the type C mussel over the brink to death.

In *Mytilus*, then, we can see the interactivity of the different steps of The Living Staircase. Selection acts upon individual mussels—these are what live or die—but it selects the winners and losers on the basis of enzyme efficiency and geographic

---

8. ATP is a molecule that provides the energy to power most chemical reactions in a cell. Its precise role in doing so is described in Chapter 3.
9. Osmosis is the movement of water across a membrane. Water moves from an area of low solute concentration to one of high solute concentration. Thus if the water around a mussel is more concentrated than its body fluids, the animal faces the danger of dehydration.

location. And the result of all of this can be observed at the molecular, cellular, organismal, and population levels.

Could selection take things this far in a computer? Could electronic viruses evolve into well adapted forms of life? To answer, we have to focus on what has been lacking in computer viruses up to the present. They respond to selection, but do not evolve (unless helped) for one good reason—an absence of mutation, which is criterion *a* for selection. When the Core Wars were in action, selection culled out the losers; but the winner had no chance of self-improvement because it did not mutate, and thus could not adapt to the changing opportunities around itself.

Today, a number of computer scientists have begun to write programs that mutate at regular intervals, with the goal of synthesizing a new form of life. Released into a confined operating environment, such programs evolve into rich assemblages that mimic biological ecosystems. Parasites show up and tap into the replication code of other "organisms," and hosts evolve immunity to their parasites. Such viruses have acquired the capacity to evolve new adaptations, and thus meet another of the defining characteristics of life.

Charles Darwin, the genius that formulated the concept of selection and recognized its power, closed his life's major work with this statement:

There is grandeur in this view of life, with its several powers, having been originally breathed by the Creator into a few forms or into one; . . . from so simple a beginning endless forms most beautiful and most wonderful have been, and are being, evolved.

I wonder if, twenty years from now, we will turn on our computers to marvel at the new "most beautiful and most wonderful" creatures.

Q & A with Jeff & Jen

JEFF: Okay, Jen, I think I've got the concept nailed down. Mutation produces variety in a population, and selection picks and chooses to favor certain varieties over others.

JEN: So far you're doing well. A modern biologist would use the term "phenotype" instead of "variety," but close enough. Keep going.

JEFF: And since selection favors a certain phenotype in location A and a different phenotype in location B, selection produces diversity in nature. Like peppered and melanistic forms of that moth.

JEN: You're right on the mark.

JEFF: Okay, so let me throw you a curve. Somewhere in the chapter you referred to selection as the force that created the amazing complexity of The Living Staircase.

JEN: Guilty as charged. What's the problem?

JEFF: I can understand how selection creates the distribution of mussels in Long Island Sound, but I can't see how it creates really complex things.

JEN: Such as?

JEFF: I always think of the eye in this case. I mean, it really is an amazing organ. There is a layer of cells that's sensitive to light intensity and color, a layer of blood vessels to nourish that layer of cells, a lens to focus the light rays, muscles to pull the lens into the right shape, other muscles to adjust how much light gets in, etc., etc. How do you get something like that by mutation?

JEN: I see the problem. Just for the record, you don't have any problem seeing how selection would favor the eye once it evolved, right? But you can't see how it could get there in the first place, by random mutation.

JEFF: Exactly.

JEN: Jeff, you've hit upon one of the oldest criticisms of the theory of evolution. Darwin had to contend with this question himself, and he devoted some time to it in *The Origin of Species*.

JEFF: So what'd he say?

JEN: He said you're right. It's ludicrous to think that the human eye could suddenly arise by mutation. But you're thinking only of the finished product. It could be that they eye started out as a very simple structure, and then was modified by a series of small improvements. Thousands, maybe millions of mutations were favored by selection, and over time there was a gradual evolution of the structure that you see today.

JEFF: That's still crazy. First, it would take forever, and second, how did we survive with primitive eyes?

JEN: Well, all evidence points to life having on earth for about 3 billion years. That's enough time to accumulate millions of mutations. And as for your second point, you're still focused too narrowly on humans. Imagine that we go back to the beginning of life, and we have a simple organism that has no eyes at all. It then has a mutation that provides a primitive light-sensitive spot—nothing as complex as a retina with a lens, but just a simple patch of cells that can distinguish light from dark. In a population where some of the animals have eyespots and the others are blind, is it hard to imagine that those with eyespots would be favored?

JEFF: I suppose not. They could tell whether they were hidden from predators or out in the open.

JEN: Okay. Now let's add on a second mutation. This one enables the eyespot to respond to the intensity of the light—so now the organism can distinguish bright light from dim light, and moderate darkness from total darkness. Would the modification be favored?

JEFF: Yeah, I get the idea. Now it can tell time of day.

JEN: The next mutation allows it to distinguish a few colors, and the next to focus most of the light on the most sensitive part of the eyespot, etc. The point is—and this is what Darwin said—as long as each step is an improvement over the previous phenotype, the evolution of complex structures is not a problem for selection theory. We have these amazing eyes because eyes have been evolving for half a billion years.

JEFF: According to what you've said, though, selection should also produce diversity. Are there a bunch of different types of eyes?

JEN: Yes, in fact. The simple eyespots I mentioned are still present in some species of single-celled eukaryotes. Insect eyes are very different in structure from our eyes, and in fact there are some mollusks that have eyes that are better designed than ours.

JEFF: What? I though mollusks were a lower form of life.

JEN: Most biologists wouldn't put it that way. It's more accurate to say they're a more ancient form of life. But "lower"? Heck, they're pretty well adapted to their environments. They have different eyes because their ancestors had different mutations than ours did. Our eyes have a sensitive layer, the retina, that responds to light. And the retina is connected to millions of nerves. The weird thing is, the nerves are on top of the retina, so light has to pass through them before it gets to the sensors. In a mollusk, the nerves are behind the retina.

JEFF: So mollusk eyes look better. Wow.

JEN: Yeah. Wow. That's why I went into biology.

RELATED RESOURCES

Broken symmetry in the genetic code?, I. Stewart, *New Scientist* 141 (March), 16, 1994.

Expanding the genetic alphabet, I. Amato, *Science News* 137 (February 10), 88–09, 1990.

Computers Under Attack: Intruders, Worms, and Viruses, P. Denning, ACM Press, New York, 1990.

Rapid turnover of plasma virions and CD4 lymphocytes in HIV-1 infection, D. Ho et al., *Nature* 373, 123–126, 1995.

Ordeal by Hunger: the Story of the Donner Party, G. Stewart, Jr., Holt and Company, NY, 1936.

The Story of Pollination, B. Meeuse, Ronald Press, New York, 1961.

The adaptive importance of genetic variation, R. Koehn and T. Hilbish, *American Scientist* 75, 134–141, 1987.

The Blind Watchmaker, R. Dawkins, W.W. Norton & Co., New York, 1987.

ROCK

Theme

Molecules, tissues, organs, and systems are "saved" in diverse, but evolutionarily related, organisms. Structures and functions at each of these levels of The Living Staircase are remarkably similar. Often similar processes use similar structures, and these similar structures are composed of and regulated by proteins encoded by genes with similar sequences.

Examples

- Common ancestry
- Homology of structure and function
- Circulation system development
- Basics of hormone action, signal transduction
- DNA and amino acids

The Concert: Guitars and Homology

I know you're thinking: "What are you, a music critic, doing writing a chapter on similarity and homology?" and "What is homology, anyway?" Don't worry; I'll get to those questions in a minute. First let me describe the scene around me, because it's quite remarkable.

I'm standing here in a crowd of 20,000 people shrieking at the top of their lungs. We're in a football stadium, it's two in the morning, and the people are yelling at an empty stage. If that's not strange enough, these folks are screaming for an encore from a band that had their first concert before I was even born. The Rolling Stones have just finished their second encore, and the people want more still. Mick Jagger is more than 50 years old, and he's still putting on incredible, flashy, mind-blowing, high-energy shows. The people around me range in age from 15 to 65, and I can't tell who's screaming louder. One guy next to me, who must be about 50, is standing on his chair *on his hands* and holding up a sign that says "Stones Forever" with his feet.

Now, what does any of this have to do with biology, or homology for that matter? Well, it all started when a biologist friend of mine, Ilka Sierra, and I were discussing how the advent of the electric guitar had revolutionized rock 'n' roll and ushered in the electronic instrument age of music. This age was instrumental, so to speak, in the development of not just rock, but also blues, rap, disco, punk, hip-hop, heavy metal, and just about all other styles of modern music. I was working on an article about the best guitar players who ever lived, and was explaining how Eddie Durham's recording with Benny Goodman in the '30's had first brought the electric guitar to prominence, before T-Bone Walker and others took over.

"This idea of the guitar and how it's evolved throughout music history reminds me of homology in biology," started Ilka. She was always comparing what appeared to me to be entirely unrelated topics to her work. I think she studies bacteria or something like that, which at the time seemed sort of useless to me. But I really like Ilka. I guess it's because we both like the same kinds of music, and when we'd met ten years ago in college, we considered ourselves the only two cool people in our entire dormitory.

"What's homology?" I asked.

"It's like similarity between things. But it's more than that; the word implies that the similarity arose for evolutionary reasons," she said. "Let me think of an example. All mammals, you, me, your dog Dave, dolphins, we all have hair. Hair is a similar, homologous structure in us all."

"A hom-olo-gous structure," I repeated slowly, "and what does this have to do with evolution or *guitars*?"

"Well, evolution and its central tenet, natural selection, are discussed in depth in Chapter 5. Remember from that chapter the example of HIV, a virus that mutates so rapidly our immune systems can't keep up with it? With so many different versions of HIV occurring at once in one person, the chances are increased that one of the viruses will be resistant to immune defenses and will have just the right traits to survive. Each new version of the virus is homologous, in fact nearly identical to its precursor; but the one that survives is, by chance, just a little bit different in a way that gives it an edge," said Ilka.

"Survival of the fittest, right?" I offered proudly, hoping at least to demonstrate some knowledge on the subject.

"Right. Homologous refers to the stuff that natural selection saves or keeps because it works," Ilka proclaimed, clearly trying her best to bring things down to a music critic's level.

"So in the guitar analogy," I tried, "the electric guitar is like an evolved version of the acoustic guitar. Their shape and strings are homologous, but one plugs in and one doesn't. They're homologous, but also different."

"Yeah. By definition, natural selection results in both similarity and diversity. The viruses in that population of HIV I mentioned are similar but different, and obviously dolphins and us are different, but we both have body hair and other similarities. And, importantly, according to the theory of evolution, both dolphins and us, as hard as it may be to believe, evolved from one common mammalian ancestor. In the same way, it's thought that guitars evolved from one ancient stringed instrument. Over the eons, new species related to our dolphin-human ancestor repeatedly branched off resulting in new species with various traits being similar or different depending on the environment.[1] This happened over and over again—eventually, incredibly, resulting in human and dolphin—and, it is still happening." Ilka went on, "But it's important to realize, as I alluded to earlier, that just because two things are similar, it doesn't mean they're homologous, that they evolved from a common ancestor. For example, the wings of bats and insects are similar, but they are not homologous."

"I think I'm getting it, Ilk. And another parallel," I said, warming to the subject, "is that both acoustic and electric guitars still exist, just like humans and dolphins do. It's not like one is 'better' than the other simply because they share a common

---

1. Chapter 7 discusses the incredible diversity within and among species that arises from natural selection.

ancestor, but each is better in its particular environment." I was doing something I'd never done—getting excited about biology. "Heavy metal would never be heavy metal without the electric guitar, and dolphins can't play either electric or acoustic guitar," I laughed.

The Development of Rock 'n' Roll and Organisms

That conversation occurred a few months ago, and ever since I've been reading more and more about homology and biology and noticing more and more about the similarities in life around me. So that's how I wound up here in the middle of the night thinking about the history of rock 'n' roll and the roots of the Stones' music, instead of jumping up and down on my seat like everyone else.

Like all species of music, rock 'n' roll has many musical roots. Its origins arise from the mixture of two cultures, black and white, beginning in the early 1900's. Black culture carried with it the spirituals and blues that had evolved from the African music brought to America by slaves. White culture brought the folk and country music of Appalachia, the Ozarks and the Southwest.

What was homologous, what shared a common ancestry, about these diverse musical styles that came together gradually and peaked in the late '50's and early '60's to form the music we know as rock 'n' roll? Well, what are the central components of rock? The Stones are giving me a chance to experience these loud and clear, as they've emerged for a final encore, "You Can't Always Get What You Want," much to the delight of the raving audience.

At the heart of rock are the driving, percussive beat, the basic chord progressions, the repetition, the central chorus with surrounding simple verses, the sometimes call-and-response of the songs, getting the audience involved. These characteristics are homologous, to varying degrees, from Chuck Berry and Bo Diddley to Jimi Hendrix and Janis Joplin, from James Brown and Otis Redding to the Beatles and the Stones, from Bruce Springsteen and Lou Reed to the Clash and Sid Vicious. And, of course, rock's cousins, like new wave and rap, share some of these homologies.

You could line up all these different musicians and their styles and see which styles and sounds one got from the other and you could identify the homologies among the processes they use to make music. For example, most use guitars to drive the music, with the bass and drums providing underlying rhythm. Most use vocals to enhance and carry the music. In much the same way as one can carry out this musical analysis, I discovered in my reading that one can also line up organisms from a diverse range of species that are going through the developmental process from fertilization to adult and see very similar processes going on, using very similar structures. It's really quite dramatic.

Watching the first several cell divisions after fertilization of a sea urchin, chicken, or human, it is almost impossible to tell which is which! Was I ever amazed the first time I compared the processes in these diverse organisms. The more closely related the organisms are in evolutionary time, the later in development you have to go to see differences between developing embryos. For example, a trout and a bass look nearly identical throughout development, only becoming distinctly different very near adulthood; whereas, developing humans and sea urchins begin to look significantly different in the early embryonic stages.

I read up on vertebrates and their circulatory systems in particular. Of course, systems start to form much later in development than the single cells I was discussing above. The circulatory system is the first to function in an organism-to-be. This makes sense, since almost anything else that is going to happen in development requires a good transportation system. The pattern of vessels that make up the early embryo's circulating system is pretty much the same from amphioxus, one of the most primitive vertebrates, to fish to man. Look at the shark and human embryonic circuits below and see just how similar they are (Figure 6.1).

Imagine staging a huge festival and inviting all the ancestors of the Rolling Stones' music. As each band plays, you select the sound, instruments, theatrics, etc. that eventually show up in the Stones' shows. Such a festival would be like condensing generations of musical development into a few hours. Well, watching a mammalian heart develop is like watching millions of years of evolution condensed into a few months. Before the developing heart of a mammal folds into its chambers, it is a straight tube—a tube that looks strikingly similar to the "hearts" of primitive vertebrates. These organisms possess a single vessel, which is homologous to this earliest embryonic heart structure in mammals.

You're probably beginning to see why I really got into this homology thing. It's practically a hobby of mine now. Call me a homologist. The heart story gets even richer. Fish hearts are only slightly different than those of primitive vertebrates. As they develop, the hearts of amphibians and reptiles, on the other hand, pass through this intermediate stage, but then add on a new stage—they pump streams of both deoxygenated and oxygenated blood that sometimes mix. The hearts of birds and mammals, on the other hand, develop to a point where they pump a stream of deoxygenated blood and a completely separate stream of oxygenated blood. Birds and mammals have folded, four-chambered hearts.

Homology is so common that we take it for granted in many fields. For example, functional and structural homology is a basic assumption in modern medicine. Surgeons sometimes depend on strong similarities between systems in different organisms. Otherwise, they wouldn't dream of performing operations such as the replace-

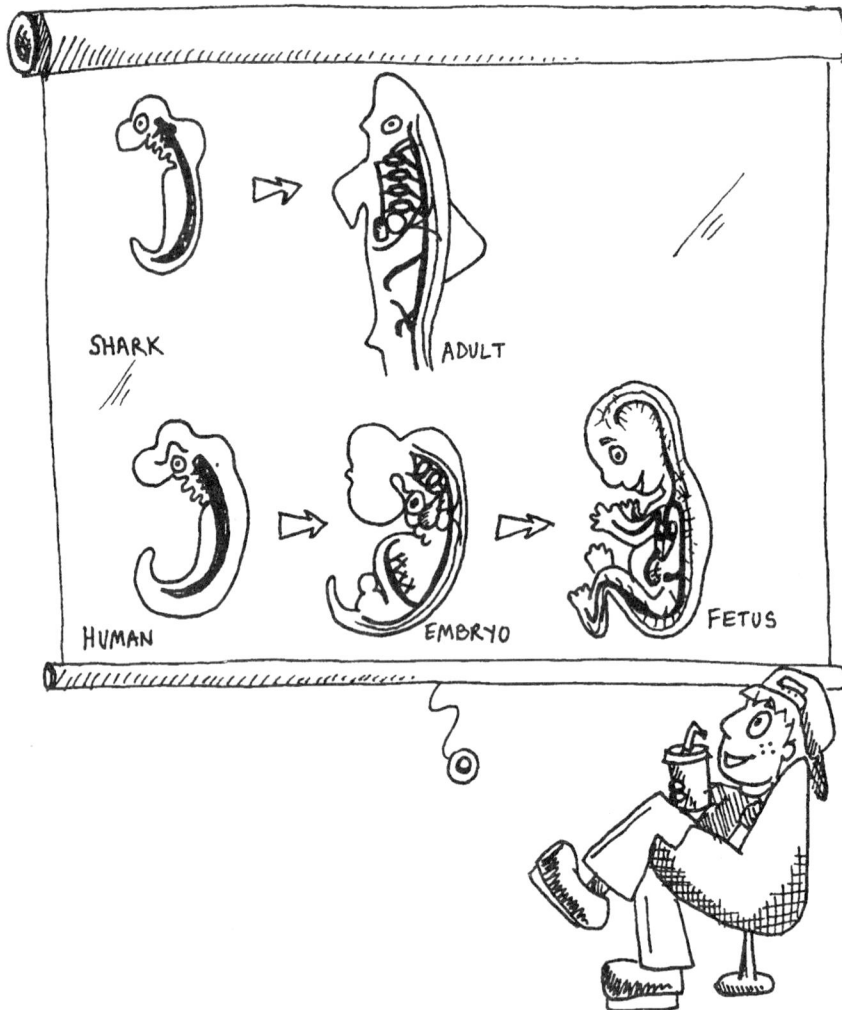

Figure 6.1 Watching the human circulatory system during development is like watching a time-lapse movie of millions of years of evolution. For example, the human circulatory system early in embryonic development looks strikingly similar to the fully-developed shark circulatory system. Evolution has "saved" basic structures and strategies for circulation (and other systems) and then added refinements for different organisms.

ment of a human's faulty heart valve with a pig's healthy one. Surgeons perform such operations routinely.

Why does the developing mammalian heart go through progressive stages of evolution before reaching its final four-chambered state? Why not go straight to the final product? Apparently, mammals have "saved" many of the genes used in the development of our evolutionary ancestors' hearts. These genes are used during mammalian heart development, and each heart is successively "expanded on" when the genes for the next "new" heart are turned on. So, in watching a human heart develop, not only do you first see an amphioxus-like heart, then a reptile-like one, then a fish-like one, and finally a mammalian version, but also genes similar to heart-building genes in all those organisms are used in the human to create each ancestral version. Pretty incredible, no?

My mind has really been wandering. I almost missed Mick and colleagues disappear from the stage. Those guys are amazing. I'll get back to this homology story as soon as I can find my car, and get home.

Okay, so I forgot I hadn't taken my car. It's been a long night. Now I'm on the subway with hundreds of other deranged rock 'n' rollers. Everyone's raucously reliving the concert. A great rock show is like a sports championship or something. It brings people together; people who wouldn't have looked at each other are suddenly best friends, slapping each other on the back and sharing food and drink. Different people suddenly become the same. Nice to see.

Sorry. I told you I was hooked on this similarity thing. Did you notice when I last left you that we had made sort of a leap in our homology discussion? It concerns that last bit about the **homology of the process of circulation system development**. The thing is: often homologous processes use homologous structures, and these homologous structures are composed of and regulated by proteins encoded by homologous genes.

Okay, I'll slow down. The central idea here is that homology at one level of The Living Staircase implies homology at other levels. Like, if you hear two bands, say Hammer's and LL Cool J's or The Beatles and The Rolling Stones, and their music is somewhat similar, then their instrumentation, their influences, and their origins are probably similar, too. Similarity often implies homology.

So, when we talk about two similar processes which employ similar structures and functions, we're suggesting that the genes responsible for those processes and the structures and functions necessary for them are homologous as well. Analogous to the rhythms and tunes of music are the proteins and biological processes of organisms. All music from ancient tribal to new age has different rhythms that are incorporated to make a tune, and all organisms are composed of and run by different proteins that work together to create different biological processes. Proteins are the molecules that do the work and give structure to organisms, organs, and cells, and genes are made of the molecules of DNA, the inherited material, that encode the proteins.[2]

Lunch and Gene Homology

The next day was Sunday, and Ilka and I had a late lunch. I related my experiences of the previous night and how I'd managed to connect the Rolling Stones and homology, even beyond her original electric guitar analogy. She was impressed.

I told her how far I'd gotten in my thoughts on homologous processes and their homologous genes, and asked her for some help in thinking of a good example.

"Getting back to your example of the process of circulatory system development," Ilka began, "what kinds of 'generic proteins' can you think of that a developing system, any system, might need?"

"Well, the analogous question might be: What types of rhythm are key to most rock 'n' roll? As far as the proteins go . . . I would guess 'information proteins.' You know, ones that let the cells know what's going on in their environments. So they know what to do next and how fast—stuff like that," I ventured.

Ilka looked proud, as if she was thinking "Could it be true? *Could* a music critic actually make something of himself?" She said, "That's exactly what researchers have found. Your so-called 'information proteins' and the genes that encode them are very similar—from organism to organism and *within* a single organism."

"Say what?" I said, "I'm lost, I think."

Our orders arrived.

"Here's an example," Ilka said around some marmalade. "The cells in an early embryo, like those in Figure 6.1, must respond to different signals, depending on what job they have.[3] The spinal cord cells have to find other spinal cord cells and move to form the proper shape of a spine, the nerve cells must find their way to the proper connections.[4] Well, we find that, in receiving their information from the environment, cells often employ molecules called G-proteins. They're named G-proteins because they interact with a molecule called guanosine triphosphate or GTP; but that's not so important for our discussion."

"How do G-proteins work? I mean, what are they?" I asked, my food growing cold.

---

2. See Chapters 3 and 4 on the processes of transcription and translation that make protein from DNA and Chapter 5 on how changes in DNA may affect the resulting proteins.
3. See discussion of hormones, receptors, and signal transduction in Chapter 9.
4. The development of neurons is also a problem involving communication, and it is discussed in Chapter 9.

"They're one of those 'information proteins' you postulated; they're sort of middlemen when it comes to getting information from outside a cell to inside. The receptors interact with signals outside the cell and then pass the signal into the cell.[5] Well, G-proteins reside partly in the cell membrane and partly in the cytoplasm and, so, they are often next in line to interact with the receptor and pass the signal on to other cytoplasmic factors (Figure 6.2). All cells need to do this environmental sensing, including our developing circulatory system cells. An organism's cells are unable to respond to anything unless they 'find out' what's going on in the environment. So, there are smell, taste, and sight G-proteins in the membranes of cells involved in those processes. And last night at the Stones concert, G-proteins were helping pass the adrenaline signal into your cells, so they could respond accordingly. So . . ."

"Gee, let me guess," I interrupted, "G-proteins are superbly homologous."

"Exactly. In fact if you look at this chart here (Figure 6.3), you can see that the amino acid sequences, the information encoded by DNA, for some of the more than 200 G-proteins so far discovered in organisms ranging from my single-celled prokaryotic favorites, bacteria, to humans are, as you say, 'superbly homologous.' In addition, the G-proteins *within* one organism, like those involved in our five senses, are also very similar to each other."

"Absolutely incredible," I said a bit too loud, causing people to glance over worriedly from surrounding tables, "it's like pieces of tunes! Themes! Used over and over again. Hey, Ilka, how come you never told me about this stuff before?" I sipped some orange juice excitedly.

"I think you sort of had to come to it on your own," said Ilka. "You know, the first thing biologists do these days when they find a new protein is compare its sequence to all other known proteins. If they can find similar sequence between a protein of known function and their own, they can be pretty sure their 'new' protein has a similar function to the known one."

"Ah, the sheer beauty of it," I exclaimed satisfied and totally carried away.

"It gets better," continued Ilka. "Scientists figured that if two proteins' sequences were similar, perhaps the proteins could functionally replace each other, and they can! For example, molecular biologists have designed yeast cells that are similar to those used to raise the dough in your biscuit there, except that they are missing one essential G-protein. Similar G-proteins, isolated from other organisms and placed in the yeast

---

5. Getting messages from the outside to the inside of the cell is also discussed in Chapter 9 in terms of vision.

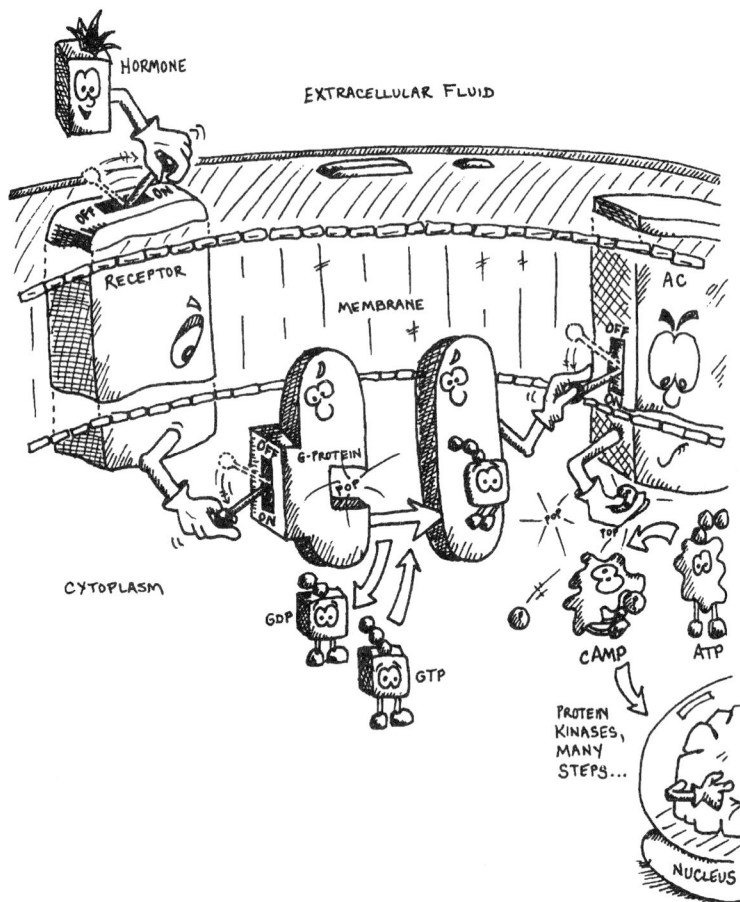

Figure 6.2 GTP-binding proteins, or G-proteins for short, help send signals from the extracellular to the intracellular environment. This is extremely important, since the environment outside the cell is always changing, and the cell must adapt to its environment to survive and perform its proper function. This diagram is of a typical G-protein system; in reality, specific signaling pathways exist for each type of signal. "First messengers," such as hormones, bind to specific receptor proteins that span the cell membrane. Such proteins are the first players in communicating information from outside a cell to the inside. (See Chapter 9 for more on how the signal "sensed" by the receptor is passed into the cell.) The receptor has three basic parts: one to contact the extracellular signal (shown here as an on/off switch outside the cell), one to span the plasma membrane, and another to send the message on inside the cell. (Other membrane-bound compartments, such as the nucleus, have similar receptors.) Binding an extracellular molecule (in this case a hormone) changes the shape of the receptor, in turn changing the shape of the G-protein bound to it, causing the G-protein to bind GTP, eventually setting off several cascades of reactions inside the cell, altering other proteins in the cytoplasm or signaling the nucleus to make new proteins. Here the G-protein is activating the enzyme adenylate cyclase (AC), which catalyzes the production of cAMP from ATP. cAMP is a "second messenger" that now signals other molecules inside the cell to turn on or off.

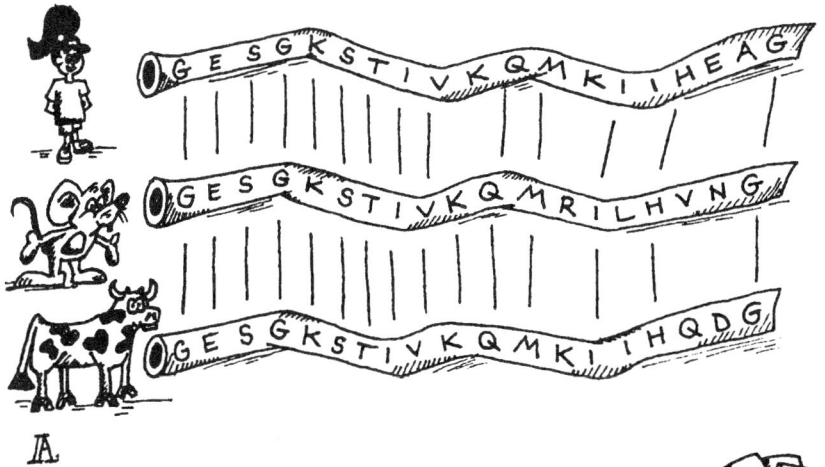

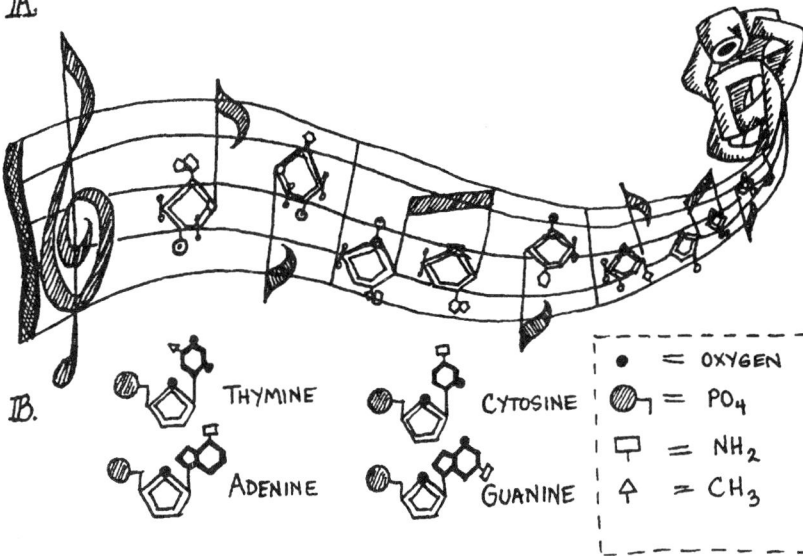

Figure 6.3 (A) Proteins, too, have been "saved" throughout evolution. Above are aligned amino acid sequences of two homologous sections of G-proteins from humans, rats, and cows. The lines connect the identical amino acids in identical positions in the three proteins. When such incredible similarity of molecules occurs in organisms whose last common ancestor lived millions of years ago, scientists figure that these parts of the molecules are the same for a reason: they must encode protein parts that have important functions. The single-letter amino acid code is: A=alanine, C=cysteine, D=aspartic acid, E=glutamic acid, F=phenylalanine, G=glycine, H=histidine, I=isoleucine, K=lysine, L=leucine, M=methionine, N=asparagine, P=proline, Q=glutamine, R=arginine, S=serine, T=threonine, V=valine, W=tryptophan, and Y=tyrosine. (B) The chemical structures of the nucleotide bases, the "notes" of life, that compose the DNA, that encodes the mRNA, that is read to make proteins, that produce the structures of organisms, are themselves very similar—both in structure and composition. The nucleotides encode a protein, and notes "encode" a tune.

cells, are able to replace the missing function! Just like one band can replace a guitarist with one from another band."

Notes and Nucleotides

That evening, after I'd finished my review of the Stones' concert, I was relaxing on the couch, pondering my new obsession, homology, and its relation to my old obsession, music, when a beautiful metaphor hit me. I jumped up to scrawl it down before it disappeared from my mind. I think you'll like it.

Notes are the nucleotides of music. Hold on, before you leave me—give me a chance. **Nucleotides are the building blocks of DNA**, all DNA in all organisms on earth. There are only four nucleotides, the same except for their attached bases: adenine (A), thymine (T), cytosine (C), and guanine (G). Within DNA, G pairs with C and T with A to form the famous double helix of DNA. The nucleotide bases themselves look pretty similar to each other (Figure 6.3). All the music of life, all similarity (and diversity) is built from these four molecules. Similarly, nearly all music, from Bach to Jagger, from Gregorian chants to gangsta rap, is built from the same few notes; three of these notes are, coincidentally, even called A, C, and G. This was too much. This was a homologist's dream.

When I told Ilka about my master metaphor, she was suitably impressed. She noted that, in a similar way, every protein in every organism is composed of different combinations and amounts of the same 20 amino acids. The amino acids are what's encoded by the nucleotide bases.

I asked her, "If all organisms use the same DNA and amino acid molecules to build their proteins, couldn't you take the homologous protein switch experiment you talked about with G-proteins in yeast to another level? Isn't it possible that the entire process of making protein from DNA is itself homologous? Just as the *specific* process of circulatory system development and the proteins involved in carrying it out are similar in very diverse species, couldn't the *general* process of making proteins from DNA be similar across species? You know—sort of like all music is made of notes and has rhythm?"

"Yes, very nice," exclaimed Ilka. "You're getting as bad as me, comparing everything to your field. Producing proteins from DNA involves two processes, transcription results in messenger RNA (mRNA) based on the DNA code, and translation results in protein based on the mRNA. These processes are so similar in all species that, as we speak, factories full of vats of bacteria containing the human insulin gene are producing insulin for use by human diabetics! Scientists provide the bacteria with the human DNA sequence for insulin. The bacteria then use that sequence to produce a protein that human diabetics can use. By implication then all the machinery, proteins necessary for transcription, mRNA components, tRNA, ribosomes, and amino acids, used to produce the insulin protein in the bacteria must be very similar to the machinery that produces it in humans. And, further, this strongly suggests that the protein synthesis processes of humans and bacteria evolved from an ancestor that we and bacteria shared a few billion years back. Otherwise the bacteria-made insulin wouldn't work as well as the human-made stuff."

As I continue my study of homology, I've sort of rethought my ideas about Ilka's experiments on bacteria being useless. I've learned that similarity of structure and function throughout The Living Staircase means that many of the things that go wrong in humans, like cancer, growth defects, diabetes, and cystic fibrosis, also go wrong in organisms that are easier to study, like mice. And because of homology, therapy or cures for diseases in organisms like mice often work to fight the analogous diseases in humans. For ethical, moral, and financial reasons, a thousand mice and their offspring are a lot easier to study than are a thousand humans and theirs. I read about a recent example of how this approach was useful in the study of diabetes, the insulin-deficiency disease Ilka was discussing above.

Scientists understand the molecular genetic mechanisms of diabetes in mice and men. The processes involved are homologous in the two organisms. As hoped, many of the same mutations in the genes that can result in diabetes in humans also appear to cause the disease in mice. Thus, researchers have developed a "model system" for studying diabetes. This greatly enhances the chances that we will discover a cure for the disease. For this exact reason, entire scientific journals are devoted largely to articles in which researchers analyze rat and mouse model systems of diabetes.

Well, since that first discussion with Ilka about electric guitars and evolution, I haven't stopped thinking and reading about similarity and homology. I've even woven some homology metaphors into my concert reviews. I continue to learn and expand my horizons, even dragging out my old chemistry book from college to discover another great underlying homology: nearly all molecules, DNA, RNA, proteins, sugars, you name it, are built from the same six atoms, carbon, oxygen, hydrogen, phosphorous, nitrogen, and sulfur, with some heavy metal elements, so to speak, like zinc and iron, thrown in. More fundamental similarity. Don't tell her, but I think I've become more of a homologist than even Ilka is. The other day, at dinner, when I asked her about the six atoms, she said, "Can't we talk about something else for a change?"

? Q & A with Jeff & Jen

JEN: Isn't it true that all kinds of people, including archaeologists, anthropologists, and linguists, crime fighters, and even lawyers are interested in comparing sequences of proteins like those shown in Figure 6.3?

JEFF: Yeah, it's because comparing sequences between two species gives you an idea of their relatedness in evolutionary time.

JEN: Meaning?

JEFF: Well, let me start at the beginning. Vincent Sarich and Allan Wilson were among the first to suggest that we could use sequence comparison to measure relatedness. The idea is based on a concept that falls out of Darwin's original theory of evolution, that any two species had, at some point, a common ancestor. We just saw the results of this basic precept reflected in examples of homology at all levels of The Living Staircase. That is, the theory of evolution implies that the more related species are, the more recently they "diverged," the more similar they are.

JEN: Yeah, the similarity between developing circulatory systems in fish and humans is really striking.

JEFF: True, but Sarich and Wilson pointed out that it's only based on similarity at the level of DNA or protein sequence that we can *quantitatively* compare species. Since any two species once had a common ancestor, they reasoned, then at some point they must have shared the same proteins, and thus the same protein sequences. Ever since any two given species "split off" from their common ancestor, their proteins have been changing with time, sharing less and less similarity.[6]

JEN: I think I'm beginning to see how this works, Jeff. If you know how long it takes for a change to occur in a protein sequence, and you know how many changes there are, then you can figure out how long it's been since those proteins were in the same species. I know techniques exist to determine the nucleotide sequence of DNA or the amino acid sequence of proteins, but how does one estimate how long it takes for a change to occur? Wouldn't different proteins change at different rates depending on their function and other unknown factors?

---

6. See the discussions in Chapter 5 concerning how change occurs and is maintained to create new proteins and eventually new species.

JEFF: Yeah, it does get tricky here. Often, DNA or protein sequence comparisons are combined with comparisons of other proteins from the same species as well as with other data to create these time scales, which are called "molecular clocks." For example, from the fossil record it's known that no primates existed before about 65 million years ago. So, all primates in existence must have shared a common ancestor not more than 65 million years ago. Knowing that, and comparing sequences for a particular protein in many primates, we start to get an idea of at least relative relatedness.

JEN: Doesn't sound too accurate to me.

JEFF: Well, you're right. Different scientists have come up with different molecular clocks, depending on what assumptions they are making, and the best they can hope for is to be within a million years or so. This isn't bad, considering . . . but here's where mitochondria come in.

JEN: Mitochondria? *What* are you talking about?

JEFF: Well, remember you started out mentioning anthropologists and linguists. Their interests in this subject have to do with mitochondrial DNA. As you know, mitochondria are our ATP-factories, our centers of energy metabolism,[7] and they have their own DNA. This DNA has a very different lifestyle than nuclear DNA.

JEN: That's right. It's only passed on from the mother, right? And it isn't subject to many of the processes nuclear DNA is.

JEFF: Yeah, and it's circular and there's much less of it than there is nuclear DNA. It has only 16,500 base pairs compared to the millions present in nuclear DNA. Most importantly for our discussion, its mutation rate, how often its DNA changes, is much faster than that of the nuclear DNA.

JEN: I get it! A faster mutation rate and fewer genes makes for a more accurate molecular clock!

JEFF: Exactly, Jen. Some scientists have used such mitochondrial molecular-clock analysis to figure out how related one human race is to another, and to postulate that the original mother, Eve, as they call her, was from Africa.

JEN: So mitochondrial DNA sequence comparison might help settle questions about which languages and cultures are related to which, and maybe even how related one individual is to another.

---

7. Energy metabolism is discussed in Chapter 4 in light of muscle contraction, and in Chapters 3 and 4, in light of the Krebs cycle and mitochondrial production of ATP, respectively.

JEFF: Right again. Population studies of mitochondrial DNA have demonstrated that the modern human populations of Southeast Asia are related to those in the many islands of Polynesia, strengthening the linguistic studies that suggested that the languages of these populations are similar.

In addition, as you suggest, degree of homology analysis of mitochondrial and nuclear DNA is now being used within the justice system to prove or disprove paternity or to help solve crimes. If the perpetrator leaves any cells, hair, skin, you name it, at the crime scene, DNA from those cells is compared to DNA from suspects' cells to help determine guilt or innocence.

JEN: Y'know, Jeff, homology has much more relevance to our lives than I ever realized.

RELATED RESOURCES

Structure and function of signal-transducing GTP-binding proteins, Y. Kaziro, H. Hoh, T. Kozasa, M. Nakafuku, and T. Satoh, *Annual Reviews of Biochemistry* 60, 349 –400, 1991.

Genetic clues of relatedness, A. E. Friday, J. Marks, V. Sarich, S. Jones, M. Goodman, and C. G. Sibley in *Human Evolution*, eds. S. Jones, R. Martin, and D. Pilbeam, Cambridge University Press, 295–320, 1994.

Home at last: a breakthrough in DNA identification lets Sally Kennedy bury her MIA son, *People Weekly* 46, 71, 1996.

Tracking marine turtles with genetic markers, B. W. Bowen, *BioScience* 45, 528–34, 1995.

Why is a cow like a pyramid? J. M. Diamond, *Natural History* 104, 10, 1995.

LINNAEUS

Theme

If one side of the coin of nature is similarity, then the complementary side is diversity. Three general types of diversity exist up and down The Living Staircase: diversity within a single organism, diversity among different organisms of the same population, and diversity among different species. All diversity in living organisms arises from blueprints composed of the same four building blocks, the nucleotides of DNA.

Examples

- Fertilization
- Sex determination
- Pollination
- Mating systems
- Hormone action
- DNA and protein composition

Meet Linnaeus

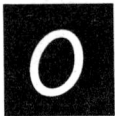

Okay, so you probably already don't like me.

Why? Because I'm the guy who came up with most of those so-called "scientific names" you're always required to learn. You know, the foreign names that are hard to spell and almost impossible to pronounce. Well, let me tell you something right up front, before you get all heated: it could've been a lot worse. Things were getting out of hand before I came along with my two-name (binomial) system. One botanist would name a flower, and then another one would come along and discover something new about it, and add that to the name. A botanist before my time, for example, called one morning-glory *Convulvus folio Altheae*; later another scientist called the same flower *Convulvus argentus Altheae folio*; in 1738, before I came up with my binomial method, even I got carried away and called it *Convulvus foliis ovatis divisis basi truncat: laciniis intermediis duplo longioribus*. You get the idea.

Ah, okay, now you're thanking me. Can you imagine learning the names of just 10 different plants with monikers like that for a test? Actually, a big part of the reason I shortened the names to two words is that my students in the field couldn't remember the extensive names I had been using.

Let me introduce myself; my name is Carl Linnaeus. I'm not nearly as old and out-dated as I sound. I'm not one of those Greek types like Augustus or Petronius, just because I happen to have an "us" on the end of my name. In fact, I changed my name to Linnaeus from Linne, just to be stylish. As a matter of fact, I'm Swedish, and I was alive when your country was born. I even had some American students. One of them became the head surgeon at a hospital in Philadelphia, Pennsylvania. I was one of the premier botanists, physicians, pathologists, geologists, zoologists, and mineralogists of my time. Coincidentally, my name actually means "lime tree." Ironic, isn't it? I'm famous for developing a classification system for plants, among other things, and I'm named after one of them.

That brings me to why the other authors chose me as ghost writer, you might say, for this chapter on diversity. I was one in a short line of folks who tried to classify nature, to catalog it into a system that made sense. Aristotle made an attempt, but for a thousand years after him, no one else even tried. When some folks finally did give it a shot, they didn't do much better than he did, if you ask me. John Ray did a decent job. He came up with the definition of species, a group of organisms which have similar anatomical characteristics and are able to interbreed, that you learn today.

Why am I talking about John Ray and his definition of species that relies on *similarity* in a chapter about diversity? Well, as you may have begun to understand in the last two chapters, The Living Staircase contains a central paradox: it is built simultaneously of great similarities and of great differences. As you read this chapter, remember the chapters on natural selection and homology,[1] and keep this paradox in mind. Many molecules, structures, and strategies are very similar in organisms that appear as different as mice and humans or plants and fruitflies. The similar-but-different paradox is integrated into every example we'll discuss. In other words, it's almost impossible to discuss diversity without mentioning similarity and vice versa.

Anyway, we were looking into John Ray's approach to cataloguing. It had problems. He and his colleagues classified organisms according to the alphabet (ridiculous, no?), and they included unicorns and mermaids in their catalogs. Looking back, I had some weird ideas of my own at the time—you folks might laugh now. For example, I didn't believe for a minute that swallows flew south for the winter, and I laughed at those who were sure that when these birds "disappeared" during the cold months, they buried themselves in the ground. I, of course, knew the real answer to the mystery: I published a paper that claimed swallows stayed at the bottom of the ocean until spring. Don't scoff: two-hundred years from now, they'll be laughing at some of the ideas you now hold near and dear.

Classifying Diversity

So, how did I improve on the classification systems of John Ray and Aristotle? Well, let me start at the beginning. When I was four I started working in my father's garden. One of the things that amazed me about the plants was how they were so different. Sure, they all had leaves, flowers, and stems, but the numbers, shapes and sizes of their leaves, and the colors and shapes of their flowers had a diversity that was overwhelming. My father was a provincial pastor and had an exceptional collection of rare plants in his garden. I bugged him until he told me as much as he knew about each one. Later in middle school, I would skip class to collect plants (yes, I was a nerd, but now I'm famous). My father was so upset when it became clear I wouldn't follow in his footsteps as a minister that he nearly yanked me out of college, until a professor convinced him I'd make a good doctor. Back then, if you were interested in any aspect of science or medicine, you went to medical school, where a broad range of sciences was taught.

Even now, as a two-hundred-and-eighty-five-year-old, the diversity of organisms has not ceased to astonish me. To this day, scientists are discovering new species on

---

1. In Chapter 5 there is a discussion of the process of natural selection and how similarity and diversity arise, and in Chapter 6 a music critic writes about the striking characteristic of similarity in nature and its implications.

a regular basis. This brings me to some other questions: why study diversity? Why spend a lifetime classifying things? You would know if you had seen the chaos in classification that existed when I was teaching and doing research. If you don't name things and organize them according to certain characteristics, you have the kinds of problems botanists were having before my binomial method became the standard. That is, when there is *no* standard, two people don't know if they're talking about the same organism or species. It's like the need to have a common language, a common vocabulary. For example, when someone published a study on *Convulvus folio Altheae*, at the time, no one could be sure that this was the same plant as *Convulvus argentum Altheae*.

In my two-name system, the first word describes the organism's genus and the second is the specific epithet of the species. For example, the Linnaean name of the romaine lettuce, *Lactuca sativa*, in your salad tells us its genus is *Lactuca* and its specific epithet is *sativa*. I further grouped species into broader collections based, again, on their similarities. Similar genera (the plural of genus) I grouped into families, families into orders, orders into classes, classes into divisions (or phyla as they're known in animal classification), and divisions into kingdoms. Our lettuce is of the family Compositae, order Campanulata, class Angiospermae, division Tracheophyta, and kingdom Plantae.

Maybe all this classifying doesn't sound like a big deal, but it is. Without a system of standardized names and fundamental characteristics to accompany them, two doctors could be talking about the same disease without even knowing it. Or worse, and this used to happen all the time to me when I was a physician, patients in the country would write me letters about a disease they had, but there were no names for diseases, except things like "Upsala [where I went to medical school] fever" or "brown death," which might refer to any number of different diseases, and there was no single set of fundamental symptoms to describe the diseases people had. Consequently, I often did not know what to prescribe. It was like a Tower of Babel.

So, I devoted my life to creating a common scientific language. It was fantastic. I or my disciples traveled all over the world collecting and cataloging plants, animals, and minerals from China to South America. During the European Age of Discovery, I placed my students on as many exploratory sea voyages as I could. Chris Tarnstrom sailed to the East Indies; Peter Kalm to North America, Fred Hasselquist to Egypt, Palestine, and Greece; and Pehr Osbeck to China. They sent me incredible collections of plant and rock specimens, the likes of which I had never even imagined. My goal was to catalog, group and name every organism on earth. I arranged all I could get my hands on into a common language called a system of taxonomy.

Being the first to organize nature into such a hierarchy was a chore. The most difficult task was thinking of names for all the organisms. I came up with over 22,000 identifying words (two for each name of more than 11,000 species). I named plants after friends, enemies, gods, goddesses, sexual organs—you name it. That reminds me of another of my biggest breakthroughs in classification: using the sexual characteristics of plants in order to classify them.

No one had done this before. Many people, including some of my esteemed colleagues, didn't believe plants even had sex, and if they did, that we certainly should not teach our young people about it. People just didn't discuss sex like they do today. One colleague said my system was bad "enough to shock female modesty." I apologize to the young ladies out there, but I want you to see what I was up against. I had to refer to the different plant classes as being like "one husband in a marriage" or "two wives in a bed with one husband." The nice thing about my system was that every plant could fit neatly into it. This was unlike any other system anyone had developed before.

Where Does Diversity Come From?

I'll get into some intriguing examples of diversity and classification soon, but before I get too far bragging about my personal accomplishments, I think we first need to address some fundamental questions about diversity, such as: Where does diversity in plant sex or in anything else come from? And, why is there such great diversity?

As I mentioned, many scoffed at my classification system based on sex in plants, but one supporter of mine was Erasmus Darwin, Charles' grandfather. I was one of the first to hint at the possibility that maybe all species weren't created simultaneously in The Beginning, that maybe some species "developed" over time from others. But I really didn't know where to go from there. It took Charles Darwin to set us straight a few years after my time. He set the stage. In his famous *The Origin of Species*, Darwin explained natural selection and, in a sense, why diversity exists.

I must say that man Darwin was a genius. His theory brilliantly explains the diversity of life. Put very simply, random mutation leads to new phenotypes, some of which by chance result in adaptation to a new environment, that is, a better chance for survival. Examples of this are the mussels and evolving viruses discussed in Chapter 5. Mutation is at the root of diversity.

Three basic layers of diversity exist in nature. Each layer is built upon the previous one. The first layer is composed of an array of short-term adjustments to the environment by single organisms; the second is a kaleidoscope of variations that occur over the very long term within a population of organisms; and the third, which also occurs over the long term, is the vast diversity that exists among different species. I first want to make sure you clearly understand each layer of diversity; later, we'll discuss examples.

Short-term diversity is built into the genes of each individual organism. The genetic makeup of an organism allows it to adjust to changes in the environment. The actual DNA of any individual organism is itself, of course, constant, but the organism's "interpretation" of its genes varies with time and situation. As I said, long-term diversity within a population arises over evolutionary time and results in real changes to the genes due to mutations that, by chance, benefit some members of the population. If you're the beneficiary of such a mutation, then you have a "greater degree of fitness," as Darwin might say, and therefore, have a greater chance of surviving and passing your genes on to a new generation.

An Example: Diversity of Sex Ratio

Let's take an example to illustrate these three types of diversity: sex ratio—the relative number of each sex to which a female gives birth. The parasitic wasp, *Nasonia vetripennis* (the two-part Linnaean name—thank you very much) provides a stunning example of the first-level of diversity. Remember that in short-term diversity, the particular environment of one individual organism determines which of several possible outcomes will occur. In this example the "particular environment" is that of the parasitic wasp's host and the "several possible outcomes" are diverse egg-laying behaviors.

Nasonia are, with ants, bees and other wasps, members of the order Hymenoptera, the class Insecta, the phylum Arthropoda, and the kingdom Animalia. I tip my hat to the good scientists who discovered the information I'm about to describe. They have taken the work of my colleagues and myself and used it as a launching pad for incredible observational endeavors.

Whether a *Nasonia* wasp develops as a male or female depends on fertilization. In order to mature into a normal adult, most organisms' eggs, humans' for example, must be fertilized. This is not the case for *Nasonia* and other Hymenoptera. If an egg is not fertilized, it develops as a male because it is haploid; that is, the egg only contains half a genetic complement. In this case the half it gets is from its mother and is in the nucleus of the egg. On the other hand, if a sperm does fertilize the egg, the diploid female develops. She has two genetic complements—one from the egg of her mother and one from the sperm of her father. So, only females get their father's DNA. The fate of a human egg's sex is radically different than this and depends on the presence or absence of the Y chromosome (see discussion below).

The female *Nasonia* lays her eggs just inside the skin of her host, a fly pupa, and her behaviors result in a diversity of sex ratios of her offspring; that is, sometimes she lays more female eggs than male, other times vice versa, and still other times the ratio is about equal. She controls each egg's sex with a special organ that can allocate a sperm to the egg or withhold a sperm from the egg prior to when it is laid (Figure 7.1).

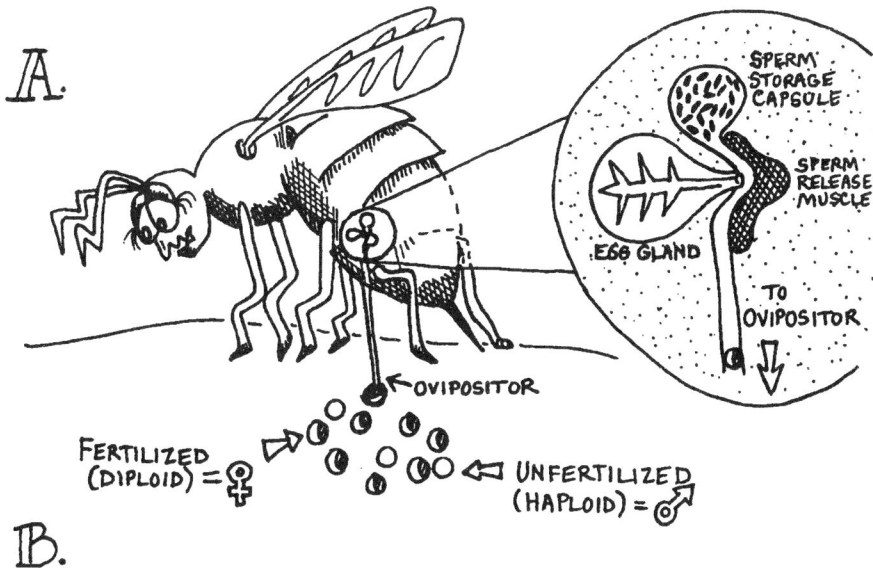

Figure 7.1 (A) Physical control of sex ratio by the female *Nasonia vetripennis* wasp. During copulation, the female receives sperm from her mate through the spermathecal duct and stores them in the capsule. When she lays eggs, she extrudes them from the gland and then, depending on the environment, she may use her sperm-release muscle to release sperm from the capsule to fertilize the egg before she lays it in the host. (B) The environment controls *Nasonia* egg-laying behavior and, therefore, the resulting sex ratio. If a *Nasonia* egg is unfertilized (haploid) it develops as a male, if it is fertilized (diploid), it develops as a female.

The sex ratio of her eggs depends on how many other *Nasonia* mothers have visited the same host and on how many are present on that host when she is (Figure 7.1). Females can tell if others have visited a host previously by detection of the venom they inject into a host prior to egg-laying. If a female is the first to find a pupa and is alone, she will lay 80% female eggs. If two or more females are present on the same host simultaneously, they each lay roughly the same number of male and female eggs. Finally, if a female is alone, but senses others have laid eggs in the host, she lays a greater number of male eggs than female eggs.

This is an example of short-term, behavioral diversity within an individual organism. Of course each female *Nasonia* has only one set of genes and only one sperm allocator, but she uses them differently depending on the situation.

There's more: nature has added some intriguing twists to the sex-ratio story in this species. A small percentage of the male wasps carry a special chromosome in their sperm, so that when a sperm containing it fertilizes an egg, this chromosome somehow destroys all the rest of the male DNA. This murderous chromosome then becomes incorporated into the genes of the egg that its sperm fertilized. This strange act of destruction results in a *Nasonia* egg that is now male because it's haploid, even though the mother "intended" for it to be diploid. Furthering the sex-ratio intrigue, there is a strain of bacteria referred to as "son-killer" which specifically kills only *Nasonia* males. This last twist gives you a hint of how studying diversity and its evolution often leads to the discovery of overlap among the evolutionary histories of very different species, in this case wasps and bacteria.

Layer Two: From One Nasonia to the Other

Aren't these wasps amazing? And to think that this is merely one species out of millions . . .

Each female *Nasonia* wasp has the ability to vary her egg-laying behavior. But as with any behavior, or any trait, for that matter, each female is different. This kind of difference among members of the same species make up the second layer of diversity. Some *Nasonia* only lay eggs on a host that is already infected. Others might have weak egg-laying muscles, hindering their laying ability, or a more efficient sperm allocator (Figure 7.1), increasing the chance they will lay a female egg. Other females have larger wings or stronger wing muscles than their relatives, so that they can fly further to find more hosts, including some not visited by other females. The females that have such traits and their offspring will also fare better in the big scheme of things.[2]

---

2. See discussion of adaptive advantage in Chapter 5.

You can see, too, how traits overlap. For example, as we discussed, egg-laying ability varies; however, any number of other traits also affect egg-laying ability. Most of these traits also differ across a spectrum. Each organism on earth is nothing more than a conglomerate of diverse traits. In turn, each species is a collection of organisms whose variation in traits fits within certain constraints. When the diversity of a group of organisms outgrows these constraints, a new species results.

For example, imagine ways that new species could develop from *Nasonia vetripennis*, the wasp we've been discussing. I'll give you a hint: pick any trait we have discussed and push it to the limit. Perhaps some *N. vetripennis* develop traits so that they can now use a new host for laying their eggs. This host is only present high up in trees. Only strong-winged wasps can reach the host. These traits, over the very long-term, affect other traits, and gradually these *Nasonia* and their succeeding generations move into the next layer of diversity, that is, they become a new species.

Layer Three: A Wealth of Variation among Different Species

We've seen the ability of an individual organism to create diversity and flexibility in sex ratio, and how there is a diversity in traits from individual to individual *within* a species. Now let's compare one trait *across* species and consider the global evolutionary diversity of sex ratio. We'll take a quick tour through the animal kingdom in order to give you a taste of the seemingly infinite diversity that exists (Table 7.1).

Here we go. Mammalian populations like us *Homo sapiens*, whose sex is determined by differences in sex chromosomes (known as heterogamety) where XX=female and XY=male, pretty much keep our sex ratio at 1:1, regardless of the environment, unless you include cultural conditions. In some cultures, one sex or the other might be "unwanted," and thus is reduced in number. Even in the apparent constancy of mammalian sex ratios, however, we find diversity. *Myopus schisticolor*, the wood lemming,[3] is heterogametic, but nevertheless also has a strikingly unequal sex ratio. Many more female wood lemmings exist than male ones, and, on top of that, there are two types of females. One kind produces only daughters, the other both sons and daughters.

In some arthropods, one gene in the mother controls the sex ratio of her offspring. *Aa* mothers produce all daughters, while *aa* mothers have all sons. In others, the father controls the ratio. He can eliminate the DNA that he donates (reminiscent of the murderous chromosome that some *Nasonia* males' sperm carry), or he can shut his DNA down during early development of the embryo he has fertilized. Sex ratio can also vary with mother's age. Some insect mothers, as they get older, give birth to more females than males, others fewer, and with others there is no difference.

---

3. See Chapter 3 for more discussion on population cycling in lemmings.

Animal	Sex RATIO	Extra...
(MOST) MAMMALS	♂ = ♀	CONTROLLED BY SEX CHROMOSOMES
WOOD LEMMINGS	♂ ◁ ♀	2 KINDS OF ♀ PRODUCE DIFFERENT SEX RATIOS IN OFFSPRING, OFTEN FAVORING ♀.
ARTHROPODS	ALL ♀ OR ALL ♂	DEPENDS ON MOTHER'S GENOTYPE
TURTLES	HIGH TEMP. → ♀ LOW TEMP. → ♂	IN TURTLES, GENERALLY COOLER TEMPERATURES PRODUCE MALES AND WARMER TEMPERATURES PRODUCE FEMALES.
ALLIGATORS	HIGH TEMP → ♂ LOW TEMP → ♀	THE OPPOSITE IS TRUE IN ALLIGATORS; BUT CROCODILES PRODUCE FEMALES AT EXTREME TEMPERATURES AND MALES AT NORMAL ONES. RESULT = BALANCE
CROCODILES	→ ♀ → ♂ → ♀	
MARINE WORMS	♂ ↔ ⚥ ↔ ♀	VARIES → ALL HERMAPHRODITES, SEQUENTIAL HERMAPHRODITES, TWO SEPARATE SEXES, OR COMBINATIONS.

Table 7.1   A diversity in sex ratio tour through the animal kingdom.

Sex ratio is at the whim of nest temperature in many reptiles. Similar to *Nasonia*, the environment mediates sex ratio, but unlike those wasps, the environment affects the eggs *after* they have already been laid, instead of *during* the laying process. Males develop at high temperatures and females at low in alligators and caimans. The situation is exactly reversed in many turtles; and in crocodiles and some other turtle species, males develop at intermediate temperature and females at either extreme.

As you can see, our tour is providing us with almost any angle on sex ratio you can imagine. The diversity in sex ratio story gets even wilder when we consider hermaphroditism, both sexes in one organism. Some species of marine worms have two distinct sexes, others are all hermaphrodites, while still others undergo sequential hermaphroditism. This last one impresses even me, the great cataloger, of whom it was said "God creates, and Linnaeus classifies." (Okay, I'm the one that said it.) But, anyway, these worms have a "pure sex-type" that does not change sex during its life and another type that starts life as the sex opposite that of the pure sex, before switching to a hermaphrodite later in life. The sex ratio in these organisms varies even within the same set of animals!

Flowers as Bridal Suites

To look at another example of diversity, I'd like to get back to plants, something a little closer to my own work, and explore pollination and pollinators. If you'll recall, I caused quite a stir in my day by classifying the great diversity of plants based on their sexual characteristics. Let's look at flowering plants (division Anthophyta, kingdom Plantae) specifically. In them, the flower is the site of plant sex (Figure 5.5). Botanists call the female part, which contains developing female gametophytes, the carpel or pistil. It is composed of the stigma, style, and ovary. The part where male gametophytes develop is called the stamen and is made up of the anther and filament. Ovules at the base of the carpels in the ovaries develop into seeds once fertilized. Pollen grains, which contain the sperm nuclei, develop in the anthers.

The generic sequence of reproductive events called pollination, most of which you can see occurring in any field in the flowering season, goes like this: pollen grains, released from anthers, one way or another get to the tips of the carpels, at which point a pollen tube grows down the style and discharges sperm, resulting in fertilization. As the embryo grows, the ovule around it becomes a seed. Remember when we discussed the paradox of simultaneous similarity and diversity? Well, in this case, I just described the constant, similar part of pollination. Now comes the diversity.

As I observed when I was a child in my father's garden, dazzling diversity exists at every step of the pollination process. The more I explored, the more I realized that almost any variation on a theme I dreamed up probably already occurred somewhere in some plant. Let's start with selfing, if you'll pardon the expression. Darwin said that nature "seems to abhor" self-pollination. We've all heard of the potential dangers of inbreeding. Plants use diverse strategies to avoid it.

Preventing self-pollination is not such a big problem if you are dioecious, that is if your female and male flowers are on different plants, but if you're monoecious (separate female and male organs on the same plant), it's another story. The problem is even greater if you have both sex organs in the same flower, known as a "perfect" flower. As a plant, you might solve this perfect flower problem by having two types of flowers: one with short stamens and long styles, say, and the other with long stamens and short styles.[4] This decreases the chance that the two gametes will come into contact. Another strategy would be to have the male and female organs operative at different times. In willow weed, stamens mature before pistils, and in some succulents, it's just the opposite.

We find an even subtler mechanism to prevent self-pollination in avocado. Each species of avocado is of two types, A and B. The A-types are receptive to pollen (that is, are "female") in the morning, and produce pollen (are male) in the afternoon. Type B is just the opposite, producing pollen early in the day and accepting it later. Thus to get avocado fruit, the product of fertilization, there must be both type A and B plants, as well as some bees to move the pollen from type to type. This is another example of diversity within a species that depends on environmental conditions; it is similar to the diverse egg-laying behaviors of *Nasonia* wasps.

We'll get to the bees and other pollinators in more detail soon. First, I want to note another variation in self-pollination avoidance, namely not avoiding it. Some flowering plants will self-pollinate if cross-pollination has not occurred, ensuring reproductive success; inbred progeny are better than none at all. Other flowering plants, such as violets, have two types of flowers—one that is relatively large and takes part in cross-pollination and another smaller type that doesn't open and fertilizes itself.

Speaking of flowers that don't open, I have to mention the case of my old pal Johann Dillenius. There's a reason you've never heard of him. He was one of those guys who spoke out against my classification by sex in plants. He claimed plants didn't even have sex and used another plant, the flowers of which don't open, to try to prove his point. Dillenius didn't realize, of course, that sex was going on—only a bit more privately. I named the plant *Ruellia clandestina*—get it, clandestine . . .? Look, you try thinking of a million Latinized names.

---

4. This is the case of the primrose described in Chapter 5 and shown in Figure 5.5.

Getting Pollinated: Variations on a Theme

Vast diversity also exists in the ways flowering plants are pollinated and who or what does the pollinating. Pollinators include bees, wasps, ants, moths, flies, gnats, bats, birds, wind, and water—just to name a few. Pollinators and pollinatees often appeared to have co-evolved in much the same way as *Nasonia* and "son-killer" bacteria did. Take the stunning example of large bees and the Scotch Broom, *Cytisus scoparius*. Only bumblebees and carpenter bees are big and strong enough to trigger the exploding broom bloom. In a young flower the keel and wings are hooked together by two teeth. The flower presents its stamens in a straight but highly tense position. When the heavy bee arrives and pushes down on the keel, the flower eventually springs open and the hidden column of shorter stamens pops out and covers the visitor with pollen. Almost simultaneously, the longer stamens and the style curve around the bee until they touch its back, the latter collecting any bee-borne pollen from previous visits to flowers, fitting the animal perfectly into a snug pollen-exchanging grip. In a field of Scotch Broom in peak pollen season, you can see bursts of bee-inspired pollen shooting off all day in a wild natural display.

The plethora of ways different flowering plants attract pollinators provides a superb example of diversity between species. Bee-pollinated flowers produce a multitudinous array of perfumes, colors and nectars to attract and guide bees to their sex organs. We've all smelled the smells and experienced the beautiful color patterns of such flowers. Flowers pollinated by birds (for example, hummingbirds, which hover around flowers they pollinate) are usually odorless, so as not to attract insects, and have no room for insects to perch. In contrast, beetles pollinate flowers like the large blossoms of magnolia trees. These flowers are without pigment and relatively odorless. My favorites are flowers pollinated by insects like flies, who are attracted to the smell of carrion. Not surprisingly, these plants smell like dead animals, which is bad enough when the plants are small. But one flower that grows in Sumatra, the giant arum lily, *Amorphophallus titanium*, which carrion beetles pollinate, smells like decaying meat and can be up to ten feet high.[5] Yes, the actual bloom is ten-feet high.

An orchid of Algeria, *Ophrys speculum*, is most spectacular in attracting pollinators. The orchid imitates the sheen and shape of a receptive female wasp, inducing, or should I say enticing, the male to copulate with the flower (and they told me plants don't have sex). Before our good wasp realizes the situation, the plant covers his abdomen with pollen which he will transfer to his next "love."

Being, as some have called me, The Prince of Botanists, I could go on forever about plants. But I'm going to discuss only one more characteristic and its diversity,

---

5. A close relative of this species, the dragon arum, produces an extraordinary amount of heat to mimic decaying animals. This adaptation is discussed in Chapter 8.

before moving on. I'll finish with the timing of blooming; an example of between-species diversity dependent on the environment. Some plants bloom in a cycle that depends on the length of darkness. Short-day plants (more properly called long-night plants) bloom when darkness is longer than a specific time. Long-day (or short-night) plants flower only if darkness is shorter than a particular period. We are accustomed to plants that bloom during the day and close at night. The Scarlet Pimpernel, *Anagallis arvensis*, does exactly the opposite. Other blooms such as those of cacti, close and open in a cycle independent of light. That is, at maturity these flowers will open and close on the same schedule even if they are kept in the dark or the light around the clock.

Intriguingly, changes in temperature can also regulate blooming time. Folks were puzzled when they could never get the tropical *Dendrobium crumentum* to bloom in greenhouses, although in the jungle it blooms beautifully and simultaneously in small localized patches. Finally, botanists determined that it was the sudden localized 10–20 °F drop in temperature that often occurs after tropical storms that induced blooming. Also, many plants will not bloom without a winter period, even though they do their actual blooming in springtime. Apple trees will bloom in Washington, but not in Florida.

Getting to the Bottom of Things

As I've learned, all this organismal diversity is rooted in variation at the cell and molecular level. Any change or variation in a behavior, for example, must be reflected in differences in the molecules controlling the behavior, the cells of the organs and the organ systems performing the behaviors, and the DNA and RNA encoding the proteins that make it all possible.

First, think about different types of cells. One basic diversification you know in sexually reproducing organisms is that between germ cells, that is the sperm and egg, and somatic cells, the rest of the cells that make up organisms. On the other hand some organisms, like bacteria, reproduce asexually and thus have no separate germ and somatic cells. Baker's yeast, *Saccharomyces cerevisiae*, can reproduce sexually or asexually—the same cell has the potential to do either. Gametes, the "sex cells," like those of *Nasonia* I discussed earlier, contain a haploid genome (half a genetic complement) that is passed on to the next generation. Clearly, gametes are very different from the somatic cells which are diploid (that is, contain two genetic complements) and make up most of an organism. Pollen grains, for example, as I discussed earlier, develop into male gametes. They attach to a stigma and grow a long projection, part of a third cell, down the style in order to fertilize the egg. The stigma and style are composed of somatic plant cells as are the leaves and roots and all other parts of the plant. Clearly, these cells, although all part of the same organism, are different in function and appearance.

Taking cell diversity one step further, there are distinct differences between the male and female gametes within one species and among other species. For example, there is a diverse collection of pollen types, which vary depending on, among other things, the medium of their transport to their female counterpart. Pollination in Mermaid Weed takes place entirely under water. This plant's pollen is composed of balloon-like cells that float in clouds above the swinging stigmas. Pollen that relies on the wind for transport, on the other hand, is light and dry and often has wings to help it be carried (Figure 7.2).

In humans, the male gametes, sperm, swim using flagella to reach the female germ cells, the ova. Sperm of the nematode *Caenorhabditis elegans* use amoeboid movements to reach their targets. Contrast gametes and somatic cells in us as we did in plants. Compare a sperm to a somatic cell, such as a liver cell, a skin cell, or a nerve cell which can be extremely long and branch off into many dendrites.[6] The great mystery in biology is how such astonishing diversity arises from cells which in a given organism *contain exactly the same genetic information*. This is related to the diversity flexibility we discussed earlier in regard to the wasp *Nasonia*. In that case it was a single organism's ability to vary, for example, its egg-laying behavior. Here I'm talking about an enormously expanded version of that diversity, that is the ability of every cell to use the same genetic information in different ways toward entirely different ends.

Can Voles Solve This Mystery?

No. But they can certainly give us a clear example of diversity of behavior and how it is mirrored down The Living Staircase to organs, cells, and the genetic information itself. What are voles anyway? My students brought me specimens of voles from all over the world. In 1758 I placed these small rodents in the Genus *Microtus* and Class Mammalia. Since then, other scientists have placed them in the Order Rodentia.

Different species of *Microtus* employ distinct **mating systems**.[7] Some species are promiscuous, that is either sex will mate with an unlimited number of the other sex. Some are promiscuous only within their own communal group. Others are polygynous, which means one male mates with many females. Another vole species, *M. ochrogaster*, is monogamous; one male and one female bond for life.

Hard to imagine two voles falling for each other at first sight, then settling down together for the long haul, but this appears to be the case. Researchers have defined monogamy in the form of two experimentally measurable behaviors: selective affiliation with one partner (partner preference) and aggression (by the male mate in the

6. See Chapter 9 for detailed discussion of the neuron.
7. See Chapter 9 for discussion of a diversity of courtship behaviors.

Figure 7.2 Pollen diversity. A taste of the astonishing diversity in shape of pollens, each with characteristics suited for its mode of transport. The pollen grains are shown as the same relative size only to make it easier to compare their shapes.

vole's case) aimed specifically at any stranger invading the couple's territory, nest, or home site. In a classic case of maneuvering along The Living Staircase, scientists have discovered the molecular cause for this animal's behavior. They isolated from the array of behavior-controlling molecules one single hormone that is largely responsible for vole monogamy.

Hormones are molecules of the endocrine system that communicate between it and other organ systems.[8] Because of information gleaned from other steps of The Staircase, the researchers suspected the hormone called arginine-vasopressin or AVP, to be a major player in vole monogamous behavior. The researchers also knew that the nervous system is the major one involved in the process of one organism recognizing another. So, the researchers compared the nervous systems of Microtus the Monogamous with those of polygamous voles. They found AVP present in both systems, but its distribution was distinctly different in the two species' nervous systems in general and specifically in certain parts of their brains. Also, when studying Microtus on its own, the scientists found an intriguing difference in AVP brain distribution between these monogamous male and female voles. This suggested an explanation for the male aggressive monogamous behavior.

Based on their discoveries, the researchers hypothesized that if they could control the release of AVP in crucial areas of vole-brains, they would be able to create monogamous behaviors in otherwise polygamous voles. In such experiments, the researchers were indeed able to directly correlate the presence of AVP to monogamy. Thus they demonstrated that a diverse behavior pattern involving the endocrine, nervous, and other organ systems was dependent on a diversity in the distribution of a single molecule—one step of The Staircase resting directly on the other.

AVP, DNA, and A, C, G, and T

The vole story gets even more intriguing if we look more closely at the structure of AVP. Arginine-vasopressin is a specific kind of hormone called a neuropeptide, which is a peptide or short protein involved in the operation of the nervous system. Amazingly, this molecule is only nine amino acids long! Nine amino acids, arranged in a particular order, Cysteine-Tyrosine-Phenylalanine-Glutamine-Asparagine-Cysteine-Proline-Arginine-Glutamate, create vole monogamy. I would bet changing a single amino acid of these nine would prevent monogamy.

Proteins are the activators and building blocks of cells, organs, organ systems, and organisms. Most proteins are built of various numbers of only 20 common amino acids in different order.[9] Yet, enormous diversity arises from these few basic molecules. If we realize proteins can contain any number of amino acids in varying order,

---

8. Hormones and communication are discussed in Chapter 9.
9. See Chapter 6 concerning amino acids and Chapter 4 for a discussion of how proteins are made.

---

from the three amino acids of thyroid stimulating hormone releasing factor to the thousands of amino acids in dystrophin, clearly there is an infinite number of possible proteins. That last one, dystrophin is the protein that when mutated results in Duchenne's muscular dystrophy, a debilitating disease resulting from the wasting away of muscle. Dystrophin is made from a region of our genome covering over *two million* base-pairs or one-thousandth of all our DNA! A mere twenty amino acids, the building blocks for this protein and all others, provide nature with vast potential diversity.

Only four nucleotide bases, adenine, guanine, cytosine, and thymine, encode these 20 amino acids and the proteins of which they are constructed. It is staggering how so much diversity explodes from only four nucleotides. And these four nucleotides are made up of only four elements, carbon, hydrogen, oxygen, and nitrogen, arranged in four variations (Figure 6.3). These four nucleotides are aligned on a backbone of carbon, hydrogen, oxygen, and phosphorus to make DNA. RNA is made of basically the same stuff, as are the amino acids themselves (of the common amino acids, two contain sulfur, and none contain phosphorus). So we've gone all the way down to the base of the matter, and found that all diversity, and in fact all similarity, rides on the back of the same few chemical elements.

This reminds us again of how the diversity, variation, and difference in nature is integrally related to the similarity and sameness found there. For example, voles are obviously different from us; however, both being mammals, we share many similar characteristics, and we also share many of the same or similar molecules. A form of vasopressin is active in us, and we are also monogamous. Maybe vasopressin is involved in monogamous behaviors in us? Maybe not. This is another example of the "simultaneous similarity and diversity in nature" paradox.

Reflections

I think back to my childhood days in my father's garden. I was inspired back then to explore and understand the incredible beauty and diversity of the plants that filled the garden: the pageantry of flowers, insects, light, shadows and color. To think that I was happy merely to classify organisms seems silly now after learning about the progress we have made over the past 200 years. But, I suppose, none of it would have been possible if it had not been for folks like me. I like to think of my cataloging as sort of like creating a dictionary of life—each organism a new word to add to the language of science, and that without that language, none of the sentences, para-graphs, and novels of today's biology could be written.

? Q & A with Jeff & Jen

JEN: Linnaeus was an extraordinary scientist. I wonder if he's aware of all the talk these days about the social and political aspects of biodiversity and its maintenance? I keep reading in the popular press about how humans are gradually destroying biodiversity and how more and more species are disappearing every day.

JEFF: You're right, Jen. The two stories I hear about most are the rapid destruction of the tropical rainforests and the battle between environmentalists and loggers over old-growth forests in the American Pacific Northwest.

JEN: Why are people so up-in-arms about extinction of species, this decrease in biodiversity? Haven't species been disappearing from earth ever since there were species around to disappear? Remember the mass extinction of dinosaurs we learned about? Bam! Thousands of species gone in a mere instant of geologic time. It seems to me the occurrence of new species and the disappearance of old ones is part of nature.

JEFF: On the surface you're correct. As one of the great writers and thinkers in biodiversity studies, E. O. Wilson, points out, almost all species that have ever lived are extinct, yet a greater number of species are alive now than ever before. He elucidates this with an analogy: nearly all humans who have lived during the past 10,000 years are dead, yet more people are alive today than at any other time during those years.

JEN: So, what do we have to worry about, Jeff?

JEFF: Hold on, I'm getting there. First, I'll address your question about species extinction being natural; then you'll begin to see the light. Extinction *is* natural. In fact, model ecosystems eventually reach an equilibrium between the evolution or introduction of new species and the extinction of others. Now comes the problem; human intervention in this natural equilibrium can greatly skew the equilibrium in the species-disappearance direction, not giving the new species anywhere near enough time to catch up. Actually, if all the species in a particular ecosystem are wiped out, no species are left to even attempt the catching up part.

Dr. Wilson uses the example of the rain forests' destruction by deforestation to give us an idea of just how far out of whack the natural rate of this equilibrium is knocked. He estimates that reduction of area in the rain forests alone has increased the rate of natural species extinction by as much as 10,000 times!

JEN: That sounds pretty bad, I admit, but humans are just another species, aren't we? What is the problem with decreasing diversity by a spotted-owl species here and a whale species there? I agree, morally, this death and destruction is repugnant, but won't nature rebound?

JEFF: Perhaps, but I think you're missing what I'm saying. You're right, a species here and there might not make a difference, nature might bounce back, but some scientists speculate that something much bigger is happening. This group suggests that species like the spotted-owl are merely warning-flags for *entire ecosystems*. Of course, if these folks are right, the conclusion is very depressing: there is little bouncing back when all habitats in an ecosystem are destroyed. Let me give you a specific example to dramatize this. In Ecuador, tropical ridges near the Andes exist like islands in an ocean. Treeless areas surround them and deep valleys separate them from neighboring ridges. One of these ridges, called Centinela, was studied by Alawyn Gentry and Callaway Dodson in 1978. Representative of such "island" ridges, it contained many diverse and unique species. Centinela had over 90 unique plant species, including some with mysterious black leaves. By 1986, Centinela had been completely cleared and was covered by the cacao crop.

JEN: That means thousands of species that haven't even been, I mean never even were, discovered are already gone.

JEFF: Now you're getting the picture, Jen. And we've only been discussing plants and animals we can see. Micro-organisms in the soil are important, of course, as are the pollinating insects Linnaeus discussed—no pollinators means no next generation of plants, which means no homes for other insects and none of the photosynthetic energy those plants would have provided. Every organism within an ecosystem is connected to every other, however indirectly. In the same way, each ecosystem, in turn, is connected to every other on earth. Plants provide the energy we need to live. Ecosystems pump gases into the air that collectively affect climates the world over. We humans, Jen, seem to be creating a dangerous imbalance in nature, don't you think?

JEN: I understand what you're saying, Jeff. I see that we are causing a rapid and extensive decline in biodiversity. But I still don't see why you're being so alarmist about a plunge in diversity. What do we really lose when we lose biodiversity?

JEFF: I don't think anyone can fully answer that question. Paul Ehrlich, another thinker and writer on the issue, uses the analogy of Russian roulette. Each species lost without serious consequences is a blank in the gun, but we have no way of knowing the consequences before we pull the trigger.

JEN: Okay, but give me something more than an analogy.

JEFF: Species, as yet undiscovered and perhaps already gone, undoubtedly would have or could have had enormous economic or medicinal benefit to humans. Two examples come to mind immediately. One kind of corn recently discovered in Mexico is unique among corn species in its perennial growth, as well as being resistant to many diseases. If this species' genes could be integrated into domestic corn, the economic potential for the world is staggering. There were only 25 acres of this corn left on earth at the time of its discovery. Another example: a rare tree that had been cut almost to extinction in the rain forests was recently found to contain an effective cancer-fighting chemical. Imagine the undiscovered potential that is destroyed with diversity.

Paul Ehrlich's predictions are even more dire. He estimates that *Homo sapiens* are currently using an enormously disproportionate amount of the earth's available energy. By including the amount of food we and our livestock eat, how much firewood and other fuels we consume, and the amount of energy we waste in crops not consumed or in the replacement of efficient ecosystems with less productive ones, Ehrlich guesses we are currently using up nearly 40% of our planet's available energy. He goes on to say that if the human population doubles in the near future, as is predicted, and if population continues to destroy habitats and diversity at the current rate, then the 80% of earth's resources we will consume and the consequences of that will accelerate the extinction of yet another species: us.

JEN: Okay, okay, I still think Ehrlich and his friends are overdoing it, but now you've got me a little nervous. Is there anything we can do to save diversity or at least improve what we have left?

JEFF: Seems like we should seriously consider answers to that question, shouldn't we, Jen?

RELATED RESOURCES

The Discoverers, a History of Man's Search to Know His World and Himself, D. J. Boorstin, Vintage Books, NY, 1985.

How fundamental are Fisherian sex ratios? J. J. Bull and E. L. Charnov, *Oxford Survey in Evolutionary Biology* 5, 96–135, 1988.

The Story of Pollination, B. J. D. Meeuse, Ronald Press Co., NY, 1961.

Fertile XX- and XY-type females in the wood lemming *Myopus schisticolor*, K. Fredga, A. Gropp, H. Winking, and F. Frank, *Nature* 261, 225–227, 1976.

A role for central vasopressin in pair bonding in monogamous prairie voles, J. T. Winslow, N. Hastings, C. S. Carter, C. R. Harbaugh, and T. R. Insel, *Nature* 365, 545–548, 1993.

The Diversity of Life, E. O. Wilson, The Belknap Press of Harvard University Press, Cambridge, Mass., 1992.

Commodity, amenity, and morality: the limits of quantification in valuing biodiversity, B. Norton in *Biodiversity*, ed. E. O. Wilson, National Academy Press, Washington, D.C., 1988.

The loss of diversity: causes and consequences, P. Ehrlich in *Biodiversity*, ed. E. O. Wilson, National Academy Press, Washington, D.C., 1988.

Hormone of monogamy, K. Fackelmann, *Science News* 144, 360–361, 1993.

chapter 8

JUNK

Theme

The Laws of Thermodynamics are universal, and thus affect every level of The Living Staircase. A continuous input of energy is necessary for life at any level. Living things transfer and transform energy in extraordinarily diverse ways, but in all cases, energy is lost as heat. This depletion of energy has consequences for the human population, as well as for all levels of The Living Staircase.

Examples

- The electron transport system
- Cell necrosis
- Tissue vascularization
- Pollination of the dragon arum
- Human metabolism
- Food webs and biomass pyramids
- Home range sizes
- The food chain in China
- The fate of the biosphere

In the fall of 1994, an advertising guru at the Milton Bradley Company hit upon a terrific idea. As a promotional gimmick for the board game "Pass the Pigs," the company initiated the National College Pigsty Search, a nationwide contest to identify the messiest dorm room in the nation. What a concept! There were thousands of entrants, of course—college students are not known for tidiness—and I regretted that the contest had not been held 15 years earlier, when I would have jumped at the chance to distinguish myself. But I wouldn't have had a chance. The winners—six roommates[1] at Case Western Reserve University—would have easily beaten the height of my slovenly living in 1979.

I am looking at a photograph of their room now. It is spacious, as dorm rooms go, but devoid of space. The coffee table and end table are littered with a giant salsa container, a slumping pumpkin left over from Halloween, a can of spray starch, a roll of paper towels, a wilting poinsettia, and several empty pop bottles. There isn't a square inch to hold anything else. Two of the roomies are sitting on a couch, pinned in by assorted junk. They look happy. Another couch is covered with piles of clothing, edging two more roommates onto the floor. They sit on either side of an empty Coca-Cola box, examining a greeting card with half-hearted interest. A giant bag of buttered popcorn and an open box of cereal lie in the background. But the most curious thing of all is in the foreground. It is a cage containing a brown and white striped . . . creature. The view is obscured by a cafeteria tray with an aged bowl of chili, complete with a fork that is cemented into the dried contents. I can see only part of the creature; it is not cat-like, and certainly not part of a dog.

In addition to collecting strange animals, the Milton Bradley champions demonstrate something I have come to call The Law of the Waif: *In absentia parentum, quisquiliae cumulant.* Translation: "In the absence of parents, junk will accumulate."

The Law of the Waif is common knowledge for anyone above the age of twenty, but most people do not realize that it arises from two natural laws, the Laws of Thermodynamics. Besides being useful to explain away a messy room, the Laws explain many of the observations that we make about nature. For example, we know that (1) a continuous source of energy is required to keep all organisms alive, (2) some tissues within an organism require greater blood flow than others, (3) herbivores in a community are more numerous than carnivores, and (4) all life on earth will eventually die out. This chapter will show that all of these seemingly unconnected facts can be explained by the Laws of Thermodynamics. We'll get to the details in a minute. Let's begin by reviewing the Laws and their general relevance to biology.

---

1. Just for the record, they were Deborah Bross, Rabia Karatela, Kiersten Lieb, Jessica Hoch, Ela Sivakova, and Julie Sietker.

The Laws of Thermodynamics

The **First Law of Thermodynamics** states that energy can be transferred or transformed, but cannot be created or destroyed. The distinction between transferring and transforming is important. A **transfer** of energy occurs when energy is moved from one location to another; a **transformation** involves a change from one type of energy to another, like the conversion of light energy into chemical energy. Money provides a good analogy. A transfer occurs when I give a dollar bill to someone else; the first law stipulates that I can move money around, but I can't create it. (Last I heard, there is a federal law to that effect as well.) I accomplish a transformation by changing a dollar into four quarters—the money is in different units, but the total value is the same. As a paper bill, or as a combination of coins, it still adds up to 100 cents.

The First Law is relevant to biology because when a cell takes in energy, it immediately transfers and transforms it. For example, most cells take in glucose whenever it is available, and they transfer the energy contained in glucose to various organelles. Before doing that, however, they transform the energy—glucose is broken down in the cytoplasm, then transported to the mitochondria, where it is converted into ATP.[2] The First Law says that a cell can take in energy and make use of it, but cannot create energy on its own.

The **Second Law** can be stated in a number of ways—open any five biology textbooks and you will see five versions of it. Putting it as simply as possible: no transfer or transformation of energy is 100% efficient. Imagine if, when you broke a dollar bill into change, you received 98 cents. This would be equivalent to 98% efficiency. Of course, the 2 cents could still be accounted for (they are in somebody's pocket), but they are no longer available for your personal use. It's the same way in a cell. When a mitochondrion transforms the chemical energy of glucose into electrical energy, it does not convert all of the chemical energy; some becomes inaccessible to the cell. The lost energy can be accounted for—it has not mysteriously disappeared—but it is in a form that the cell cannot use. It becomes heat, in fact—the heat that is keeping your body temperature at roughly 37 °C right now.

A Detailed Example: The Electron Transport System

As humans, we are one of the few organisms that use the heat of metabolism to maintain a constant body temperature. Most of this heat is produced by the activity of mitochondria, the cellular organelles that transform food energy into ATP energy. A mitochondrion is composed of two membranes, and this transformation is done

---

2. The mechanics of glucose transformation are covered in Chapter 3.

primarily by the electron transport system (ETS), which is housed in the inner membrane (Figure 8.1). The outer mitochondrial membrane transports partially digested fuel molecules from the cytoplasm to the central region of the mitochondrion, where the Krebs cycle operates.[3] There, the imported molecules are broken down into several products. The most important products for our discussion are molecules of NADH and $FADH_2$, which deliver their hydrogens (protons and electrons) to the ETS.

When the electrons first encounter the ETS, they are high in energy content; in other words, the electrons cycle at a considerable distance from the nucleus of the atom that receives them. The ETS enables the electrons to drop to lower orbitals, thereby releasing the energy they contain. A portion of the energy released becomes heat, and the rest is used to make ATP. To see how this is done, we must get into the details.

A close look at the ETS reveals that it is a collection of proteins embedded in the inner membrane (Figure 8.1). The first protein, FMN, is the one that interacts with the products of the Krebs cycle. NADH and $FADH_2$ from the Krebs cycle approach FMN and drop off their protons and electrons (i.e., the hydrogen atoms). From FMN, the protons and electrons go separate ways. The protons are shuttled to the outer compartment (the intermembranous space) of the mitochondrion, where they are trapped for a while. The electrons, however, are passed from protein to protein within the ETS.

As the electrons move from one protein to the next, they drop to lower orbitals, and thus release energy. As an analogy, imagine a student that is eating a candy bar at the top of a tall staircase. Once she finishes eating, she is supercharged—loaded with energy that can be released if there is an opportunity. By jumping step by step down the stairs, she can release the energy in small amounts. Actually, to burn off the energy in a candy bar—typically about 250 Calories[4]—she'd have to jump down miles of stairs. The ETS accomplishes the goal much faster; the electrons jump through a series of three protein complexes, and finally end joining an oxygen molecule. At that point, the energy release is complete.

Now let's pause and see how the ETS conforms to the First Law of Thermodynamics. The ETS does not create energy; its energy comes from NADH and $FADH_2$. The ETS only transfers and transforms it. It is transferred each time the electrons jump from one protein to another. The transformation, i.e., the conversion of the energy from one form to another, is more subtle. Initially, the electrons are chemical energy—hydrogens that are delivered to FMN. Once the protons and electrons separate, however, the electrons flow through the system, becoming electrical energy for

3. The Krebs cycle is described in detail in Chapter 3.
4. One Calorie (or kilocalorie, in proper biological terms) is equivalent to the amount of heat it takes to raise the temperature of 1 liter of water by 1 °C.

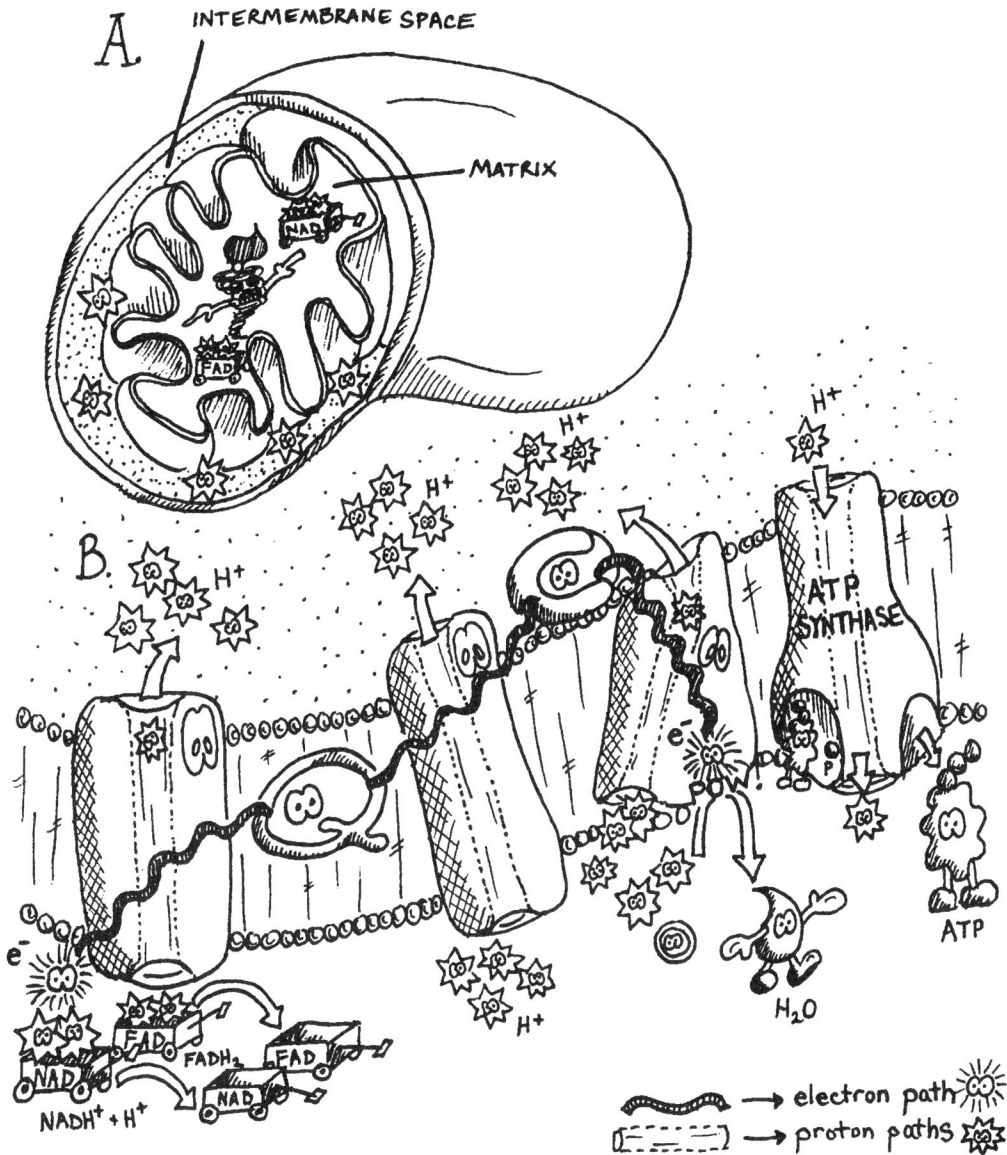

Figure 8.1 A mitochondrion is composed of two membranes that subdivide the internal spaces. (A) The outer compartment is called the intermembranous space. When the mitochondrion is active, protons accumulate there. The inner compartment, the matrix, is where the Krebs cycle produces NADH and FADH$_2$. (B) The electron transport system (ETS) is a collection of proteins built into the inner membrane. Electrons are shuttled from one protein to the next. Some of the ETS proteins also transport hydrogen ions from the matrix to the intermembranous space. The ions accumulate there, then pass through the ATP synthase complex. In doing so, they release energy that is used to make ATP.

a brief moment. This electrical energy is converted back into chemical energy later, a point we will return to in a moment.

The ETS also demonstrates the Second Law, which states that no energy transformation is 100% efficient. In other words, of the total energy released from electrons in the ETS, some is lost as heat and the remainder is used for work. But what work is the ETS doing? To answer that, we need to look more closely at the movement of protons.

Each time that the electrons are passed from one protein to another, they release energy. Some of the electron carriers in the ETS use this energy to transport protons from the mitochondrial matrix to the intermembranous space. These protons are moved **against** their concentration gradient, i.e., they are moved from an area of low proton concentration (the matrix) to one of high proton concentration (the intermembranous space). Since positive charges repel one another, it takes an input of energy to move the protons in this direction—and this is the work that the ETS accomplishes.

Once in the intermembranous space, the protons are trapped by a membrane on either side. As the ETS continues to pack in more protons, the protons build high potential energy to diffuse **down** their concentration gradient. Positioned in the inner membrane near the ETS, there is a protein complex called ATP synthase, which acts like a tunnel—it allows protons to diffuse through (Figure 8.1). As the protons move through, they cause a structural change in the tunnel that leads to the production of ATP molecules.[5] Summing up, the ETS starts with chemical energy—the hydrogens dumped by NADH and $FADH_2$—and transforms it into electrical energy. The electrical energy is used to transport protons into the intermembranous space, where they accumulate. When the protons cross back, a new form of chemical energy (ATP) is produced.

Is It Worth the Trouble?

The Second Law stipulates that energy must be lost with each of these transformations. So why do it? It seems wasteful for a cell to dismantle one form of chemical energy (glucose) just to produce another (ATP). It is worth the trouble, however. Glucose is a high-energy molecule, containing 686 kilocalories of energy/mol.[6] This is far more energy than a cell can use at once—rather like using a stick of dynamite when a firecracker will do. ATP, on the other hand, contains only 12 kcal/mol, a figure that is more appropriate for a cell's needs. Most of the reactions that a cell carries out require about this much energy. By digesting glucose, a cell repackages the energy in

---

5. ATP is the main source of energy that a cell uses to power chemical reactions. More details about its role are given in Chapter 3.
6. In other words, if you burn a sample of glucose that is equivalent in grams to its molecular weight, 686 kilocalories of heat will be released.

manageable form. A typical cell makes 36 ATPs out of each glucose molecule digested. From this, we can calculate the efficiency of the process. The 686 kcal/mol of glucose is converted into 432 kcal/mol of ATP (12 kcal/mol x 36). Thus, about 63% of the glucose energy becomes ATP, and the remaining 37% is lost as heat.

Entropy

So far we have applied the Laws of Thermodynamics to mitochondria; but how are they related to dorm rooms and The Law of the Waif? Strange as it may seem, the answer will be clearest if we return to the messiest dorm room in the nation and think of it as a gigantic cell. The analogy is not that much of a stretch, either. I'm sure that room started out as a highly organized, semi-efficient entity. Sections were compartmentalized, as in a cell, and set aside for specific functions—sleeping in one area, studying in another, eating in another.[7] Also like a cell, energy entered the room (human beings returning from a meal), some of which was lost as heat while the rest was used for work. However, there is an important technicality concerning the nature of the work: for the room to stay organized, some of the work has to be directed toward maintaining it. Otherwise, the degree of disorder will begin to rise. Physicists speak of disorder as "entropy." Using their terminology, the Case Western dorm room would be high in entropy ("It's not messy, Mom, it's entropy-rich") while a cell, being very complex and ordered, is in a low-entropy state.

An increase in entropy is a natural consequence of letting things happen as they will. The Case Western room is a perfect example—junk accumulated because nothing was done to prevent it. To hold entropy steady (stay organized) or force it to decrease (get more organized), there has to be an expenditure of energy. For a teenager's room, that expenditure of energy usually follows a direct parental command.

Like a dorm room, a cell has a natural tendency to increase in entropy over time. However, it doesn't—in fact, most cells become more organized (i.e., decrease in entropy) as they grow. This is possible only as long as two criteria are met: (1) the cell has a continuous energy input, and (2) it devotes much of this energy to maintaining and/or increasing complexity. For example, every cell has organelles called ribosomes that are involved in the synthesis of proteins. There is a constant demand for new proteins in a cell, so ribosomes are kept busy. Like a sewing machine that runs 24 hours a day, every ribosome eventually wears out. This represents an increase in entropy because a highly structured ribosome breaks down into unorganized pieces. To compensate, the nucleolus of the cell produces new ribosomes to replace the ones that have stopped functioning. Thus the increase in entropy (breakdown in ribosomes) is matched by a decrease in entropy (production of new ribosomes), keeping the

---

7. All organized systems have such a division of labor. For a discussion of its importance, meet Mr. Ira Bloomfield in Chapter 2.

overall entropy of the cell constant. When a cell is growing, it produces ribosomes at a rate that is faster than the breakdown of existing ribosomes; thus a growing cell is in a state of decreasing entropy.

The synthesis of new ribosomes can only continue as long as a cell has a source of energy to devote to it. Any time that a cell is cut off from its energy source, the struggle is lost—entropy increases. In other words, the cell begins to die. All cells, even when dormant, require energy to stay alive.

This concept has real-life applications. When the energy supply is cut off, cell degradation occurs in a predictable sequence (Figure 8.2). In the earliest stages, DNA in the nucleus clumps together, and the mitochondria begin to swell. Following this, the ribosomes disintegrate and the internal and external membranes break apart. Finally, the nucleus splits into several "ghost nuclei." By looking at the structure of cells through a microscope, a forensic scientist can determine how long entropy has been increasing, i.e., how long ago death occurred. All because of the principles of Thermodynamics!

Thermodynamics from Tissues to Organisms

We can choose virtually any level of The Living Staircase and see the principles of thermodynamics and entropy in action. Since we have already covered a detailed example at the subcellular level—the mitochondrion—and have talked in general terms about cells, we'll take another step along The Living Staircase to begin.

Tissues

In any multicellular organism, groups of cells are specialized to perform certain functions. Cells with the same function are usually positioned close to one another to form a coordinated tissue. Some tissues metabolize actively, meaning that they do a lot of work, while others are rather sluggish. By merely looking at the tissues through a microscope, you can distinguish between active and sluggish types.

The Laws of Thermodynamics dictate that active cells require a greater energy input per unit time than sluggish cells. Since the energy for metabolism comes from nutrients in the blood capillaries, there is an easy cue to rely on: active cells will have a greater degree of **vascularization**—a more intricate network of surrounding capillaries—than sluggish cells. Simply by observing a sectioned tissue, then, we can make an educated guess at its relative metabolic activity of a tissue.

For example, muscle cells have very high metabolic rates, and an extensive meshwork of capillaries surrounds them. Tissues used for storage, like the adipose layer which stores fat just under the skin, are much less vascularized (Figure 8.3). In addition, muscle cells have many more mitochondria than adipose cells. This makes

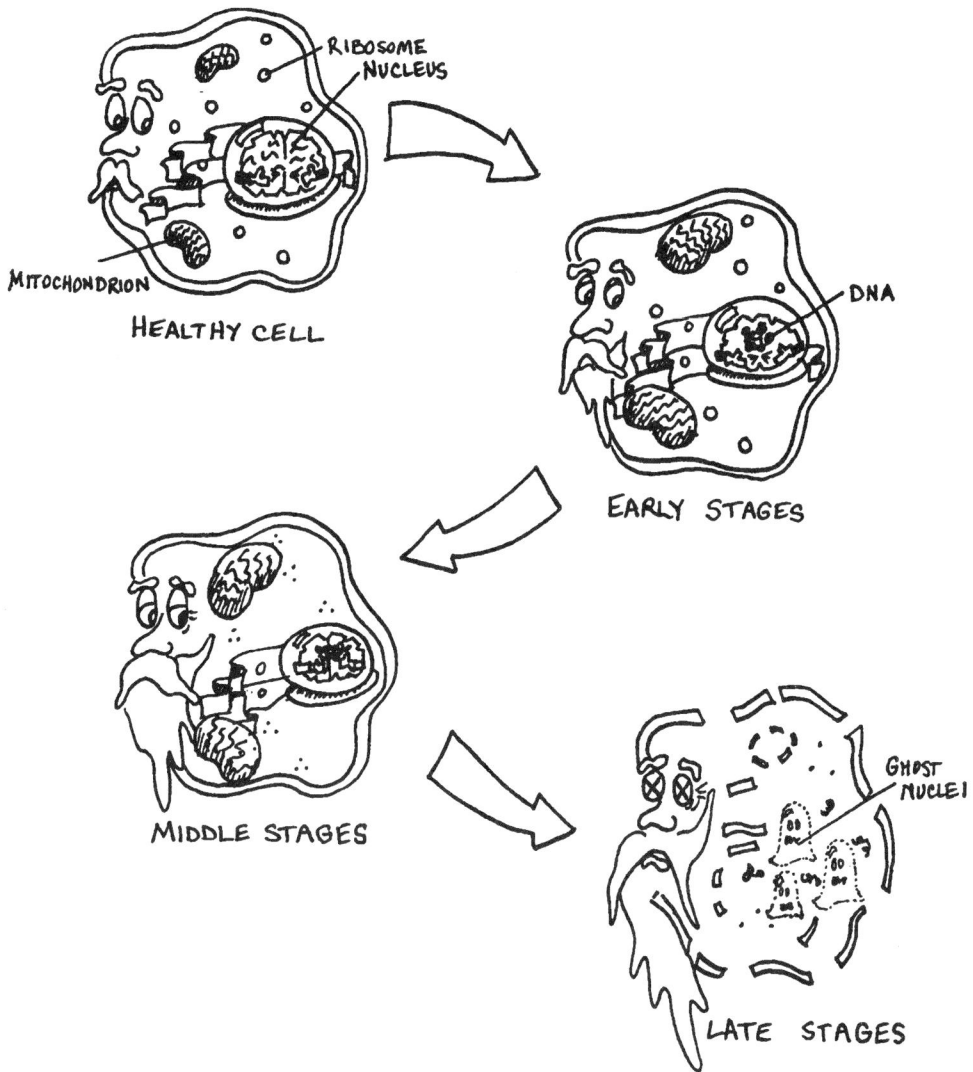

Figure 8.2 The increase in entropy of a dead cell follows a predictable sequence. First, clumps of chromatin (DNA and protein) appear in the nucleus, and the mitochondria swell. Next, the ribosomes disintegrate. In the last stages, cellular membranes break apart and the nucleus fragments into several ghost nuclei.

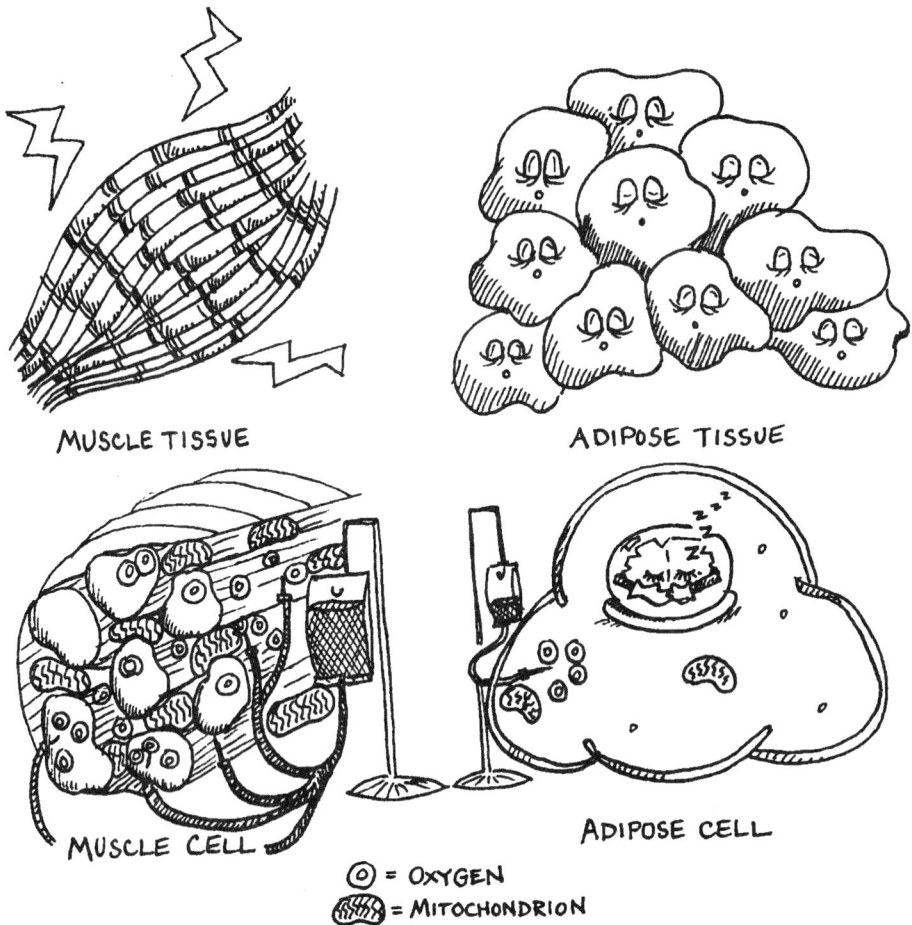

Figure 8.3 The Laws of Thermodynamics determine the degree of tissue vascularization as well as cell structure. Active cells, such as those in muscle tissues, require a high input of energy. Nutrients are brought to the cell via an extensive network of capillaries; once inside, nutrient energy is transformed into ATP energy by abundant mitochondria. Adipose cells are relatively inactive, and consequently have fewer blood vessels and mitochondria.

sense because mitochondria are necessary to turn the chemical energy of food molecules into more usable packets of ATP energy. The Laws of Thermodynamics simultaneously influence two levels of The Living Staircase: cell structure and tissue vascularization.

For tissues with very slow metabolic rates, and thus little need for energy input, death of the cell can come long after clinical death of the individual. The prizewinners for this holdout are cartilage cells, chondrocytes, that have such a low demand for energy input that there are no blood vessels passing through the matrix that houses them. Chondrocytes are known to remain intact—alive—for up to 60 years after clinical death has occurred. For that reason, it is possible to rebuild a knee using cartilage from a person that has been a cadaver for decades!

Organs

The Second Law of Thermodynamics dictates that any transformation of energy will produce heat as a waste product. For endotherms (birds and mammals), the conversion of energy to heat is obvious; they use the heat to maintain constant body temperatures. But what about other organisms, such as plants? They should also produce heat, according to the Second Law, but they aren't warm to the touch. Why not? The answer is that their metabolic rates are much lower than those of endotherms, so their heat production is low and hard to detect. When I was teaching biology labs in graduate school, we proved this by placing peas in a thermos bottle and inserting a thermometer through a hole in the cap. In a second thermos, we used peas that had been killed by cyanide. We also had a control thermos, which was empty. The living peas warmed their thermos by 1 or 2 °C, while the dead peas had no effect on temperature. For most plants, this heat loss is nothing more than wasted energy. A few species, however, have harnessed the heat for a useful purpose. The **dragon arum** is an amazing example.

A hiker that discovers this flower may be initially pleased by its beauty, but the pleasure is short-lived—the dragon arum produces an overwhelming stench which attracts flies for pollination (Figure 8.4). Unlike bees which collect nectar and pollen, flies visit a flower to lay eggs, mistakenly believing it to be a decaying carcass.

The dragon arum is a master at mimicking a decaying animal. In addition to producing the appropriate odor, it mimics the high temperatures associated with decomposition. A specialized structure, the spathe, rises from the center of the flower and generates a temperature close to 40 °C for a full 24 hours. Fooled by the combination of scent and heat, a visiting insect crawls deep inside and lays eggs. In the process of exploring the flower, it transfers pollen from the stamens to the stigmas.[8]

The central spathe of the dragon arum is a remarkable adaptation. It generates as much heat as corresponding tissues from warm-blooded animals, although the supply of energy to the flower's cells is not even close to that of an endotherm. Where does all the heat come from? An understanding of thermodynamics should lead you to expect that it arises from a markedly inefficient transfer or transformation of energy.

---

8. Pollination has fascinated biologists for more than 300 years. In Chapter 7, Carolus Linnaeus relates his enthusiasm for the subject.

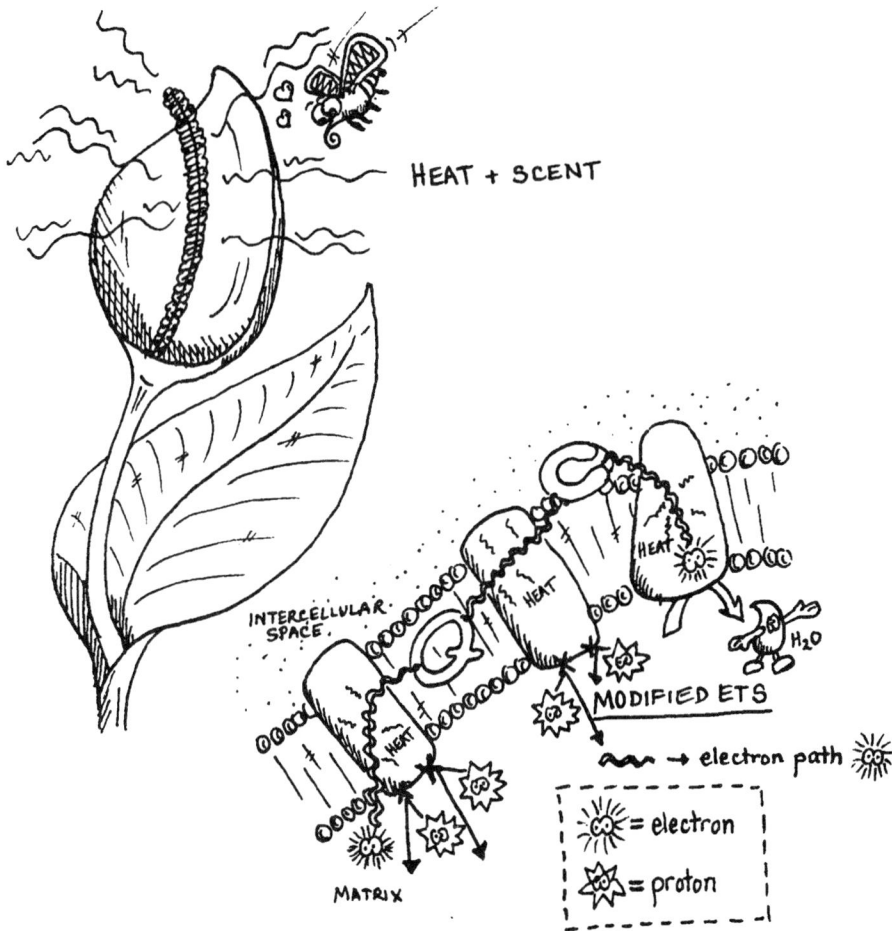

Figure 8.4 The flower of the dragon arum. (A) The central spathe generates heat that, in combination with the odor emitted by scent glands, attracts carrion flies. By crawling deep into the flower to lay eggs, the flies transfer pollen. (B) The central spathe generates heat by altering the electron transport system. In this ETS, none of the energy released from electrons is used to transport protons. Thus, all of the energy becomes heat.

("Inefficient" because heat represents energy that is lost from the cells.) To see how it is done, we have to return to the electron transport system. Remember that in the ETS, electrons are passed from one protein to another in a sequence, releasing energy at each step. In a standard ETS, some of the energy released is used to transport hydrogen ions across the inner membrane, and the rest is released as heat. In the dragon arum, however, none of the energy goes for proton transport; instead, all of it becomes heat to attract flies (Figure 8.4). This way, the spathe of the dragon arum produces heat at a rate equivalent to that of a decaying animal.

This example reveals a fundamental point of biology, and the main purpose of this book: to connect the steps of The Living Staircase. Because we typically learn about pollination in one course and mitochondria in another, the connections are easily missed. The dragon arum brings our knowledge together by linking two levels of The Living Staircase beautifully—it uses a subcellular metabolic process to achieve an organ-level adaptation.[9]

Organisms

Among animals, two groups, the mammals and birds, have evolved to make use of the heat lost by metabolism. The rest of the animals, ectotherms, also generate heat by virtue of Thermodynamics, but heat production is not a major goal of metabolism. Ectotherms have lower metabolic rates and do not maintain constant body temperatures. Endotherms maintain a high body temperature, but do not have a specially altered ETS to achieve this—they simply burn more food molecules per unit time to produce the heat. What is the advantage of generating so much heat? Apparently, endotherms benefit by being able to remain active when it's cold outside. On a cold morning, a lizard moves slowly because its body temperature dropped overnight, but a mouse wakes up ready and raring to go.

Whether an endotherm or an ectotherm, a multicellular organism is a whole-body demonstration of energy exchange, much like a cell is a microscopic example. Consider yourself. Chemical energy (food and drink) is put in, and the molecules are digested to release energy. Part of the energy is used for work, like moving your eyes across a page to follow a sentence, and the rest is the heat that maintains a nearly constant core temperature of about 38 °C. And since you are a thermodynamic unit, your inputs and outputs can be described in terms of heat.

A typical human consumes about 2000 kcal (kilocalories) of food a day. One kcal is equivalent to the amount of heat it takes to raise the temperature of 1 liter of water by 1 °C. How much of the energy put into your body is lost as heat? The answer varies according to what you are doing at any point in time. Let's start with the simplest case—a person doing nothing. One of the Case Western roomies can serve as an example. (What follows is not a criticism so much as a deduction. Let's face it—the accumulation of junk in that room would support the conclusion that they weren't doing a heck of a lot of work!)

---

9. In other words, the flower's heat production is an emergent property of mitochondrial activity. See Chapter 1 for an overview of the concept of emergent properties.

Let's imagine that one of the roommates eats 2000 kcal of food a day, and exercises only by pressing the buttons on the television remote. As energy enters her cells, it is transformed into ATP energy at an efficiency of around 50%. (The 63% figure that was cited earlier in the chapter applies specifically to the transformation of glucose into ATP. Most students eat a lot more than glucose.) Looking at it another way, we could say that 50% of the energy she takes in is lost as heat right away. Following that, the ATP molecules are broken down to power cellular reactions, and their chemical energy was also lost as heat (Figure 8.5). Unless she is doing work, virtually 100% of the energy of food is converted to heat. This is true of any adult at rest.

What if she channels some of the energy to work, like picking up books and returning them to a bookcase? In that case, some of the energy consumed will be transformed into a form other than heat. Even at peak efficiency, however, a human is able to transfer less than 25% of the energy of food into productive work, a figure which is roughly equivalent to that of a lawn mower. It is an interesting notion—we are less efficient than many of the machines we make.

What happens if she puts in more energy than she uses? For example, by consuming 2500 kcal in Cheetos while burning off only 2000 kcal? In such a case, the excess energy is stored as fat. This is an interesting characteristic of humans. Most other animals regulate the inputs of energy with the production of heat, and thus do not have a tendency to gain weight unless they are preparing for a period of time when food will not be available.[10] Bears, for example, put on fat before hibernation, but burn it off during the winter, when they don't eat. Other animals don't have the propensity for obesity that humans show. Why are we different? One explanation is that our ancestors evolved in food-scarce environments, where it was adaptive to eat a lot when food was available. In today's industrialized world, food is abundant, but we retain the propensity to eat—and thus end up heavier that we should be. The only way to lose excess fat, then, is to use more energy than is taken in. Like it or not, the Laws of Thermodynamics leave us with no alternative.

Thermodynamics and Ecology

Thus far we have looked at entropy and the Laws of Thermodynamics at levels of The Living Staircase ranging from molecules to individuals. But these principles apply to populations, communities, and ecosystems as well—in fact, one can argue that they take on increasing importance at the upper end of The Staircase. Limits to energy exchange in large part determine population sizes—and our own population

---

10. This is an application of the principle of regulation. Chapter 4 covers many examples of precise regulation, which is characteristic of biological systems.

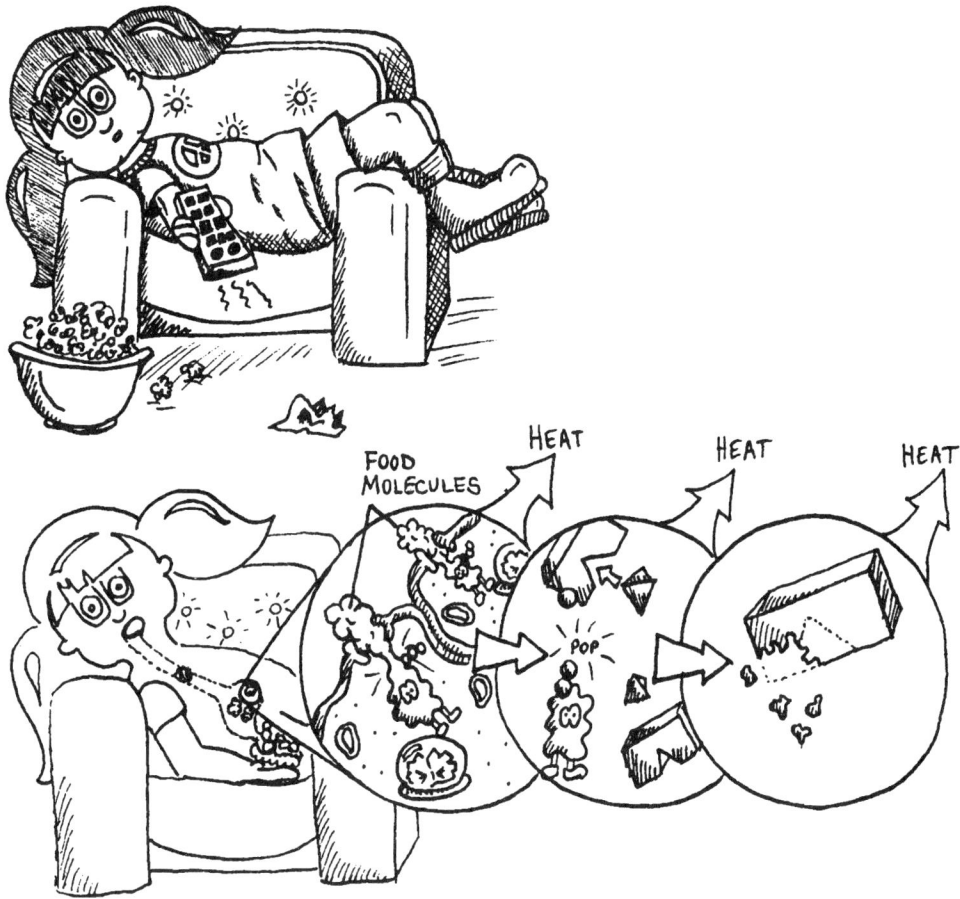

Figure 8.5 Energy transformation in a couch potato. (A) A roommate at peak performance. (B) The roomie as a thermodynamic system. Energy entered her body in the form of food, and was absorbed into her cells. Initially, the food molecules were converted to ATP, with a substantial loss of heat. When the ATP molecules were used to run cellular reactions, additional heat was lost. And when the products of these reactions degraded, all of the remaining heat was lost.

is no exception. The most intelligent species on the planet is restricted by the same laws as the rest.

As mentioned in the previous section, individual organisms can be seen as thermodynamic units—energy enters the organism and is transferred and transformed, ultimately being lost as heat. Furthermore, we can trace the transfer of energy from one organism to the next. As animals, we capture energy from other organisms. When we eat meat, for example, our energy source is that of a preexisting animal. What did the animal use as a source? Vegetation, probably. And what was the energy source for the vegetation? Sunlight.

What I have outlined in reverse order is a food chain, a central concept in ecology. Just as we can think of each individual in the chain as a thermodynamic unit, we can look at the entire chain as a unit. Energy enters at the front, when a plant transforms sunlight energy into chemical energy, and then it is transferred from one organism to the next until the end. The First Law of Thermodynamics stipulates the amount of energy moving through the food chain is set by plants; by photosynthesizing, they convert light energy into chemical energy. As this energy moves through the rest of the food chain, additional energy cannot enter. The Second Law has even more important implications; it stipulates that as energy is passed through the food chain, it will undergo a continual depletion. Why? Because each time the energy is passed from one link to the next, some of it will be lost as heat. In other words, each species in the chain has less energy available to it than the previous species.

This is reflected in biomass pyramids, which ecologists use to compare ecosystems. The species that make up a community can be divided into different groups, based on their roles in a food web.[11] The groups are referred to as the trophic levels of an ecosystem. Species that synthesize their own foods are called producers; for most communities, these are plants. Herbivores are "primary" consumers, predators that eat herbivores are "secondary" consumers, and predators that eat other predators are "tertiary" consumers. Finally, some organisms feed off the tissues of dead producers and consumers; these are called the decomposers, and are primarily bacteria and fungi.

If an ecologist is studying a community with defined boundaries, say, an acre of meadow, it is possible to determine the total biomass of each of the ecosystem components. For the producers, the total biomass would be the total dry weight of all the living plant material on the plot. What does this have to do with energy? The molecules making up the plant, other than the water, represent stored chemical energy—the energy that is available to the primary consumer. Likewise, the total biomass of the primary consumers represents the energy available to secondary consumers.

Figure 8.6 shows biomass pyramids for different ecosystems. As dictated by the Second Law, the biomass decreases dramatically from one level to the next. In other words, the biomass of predators is always smaller than the biomass of their prey.

---

11. A food web is a realistic representation of the complexity of energy flow in a community; the previous concept of a food chain, a linear transfer from one organism to the next, is an oversimplification that is useful in pointing out basic principles.

A. CORAL REEF, ENIWETOK ATOLL

B. OLD FIELD, GEORGIA

Figure 8.6 Biomass pyramids from two ecosystems. In each, the bottom block represents primary producers, the middle block is herbivores, and the top is predators. The Second Law of Thermodynamics stipulates that energy flow decreases as you move up the pyramid, but ecosystems can differ in the degree of the decline. The coral reef is more efficient at transferring energy from producers to consumers, and thus can sustain larger populations.

Notice that the pyramids differ between ecosystems, and thus ecologists speak of different ecosystems having different efficiencies. In the coral reef ecosystem of Figure 8.6, the primary producers transfer a greater amount of energy to the herbivores than in the terrestrial ecosystem. Consequently, the coral reef would be able to sustain a longer food chain (or a more diverse food web) and larger populations. In addition to determining the properties of our mitochondria, the principles of Thermodynamics affect the structure of entire ecosystems.

Biomass pyramids give us a picture of the abundance of energy at the different trophic levels of an ecosystem, and this is reflected in the behavior of the consumers. Wildlife biologists speak of individual organsims having a "home range." This is the amount of area that the animal traverses over the course of a day to gather the food it needs to survive.

Not surprisingly, home range size is strongly related to body size. A larger animal must consume more calories to sustain its metabolism, and must travel a greater distance to acquire them. In addition, herbivores and carnivores of equal body mass have quite different home ranges—typically, the carnivore's home range is about four times larger than the herbivore's. Why? Because the energy that sustains an herbivore is more abundant in an ecosystem. The carnivore's food supply (herbivores) occupies a higher trophic level, and thus is relatively scarce. The carnivore has to cover more ground to encounter it.

The fact that energy is less abundant at higher trophic levels has a strong influence on human populations as well. For the most part, we exist with a food chain of one or two links: we raise vegetation and eat it (a one-link chain) or raise vegetation, feed it to farm animals, and then eat the herbivores (two links). Some human societies exist almost exclusively on a one-link chain, while others have a diet that more often includes two links. In fact, the diet of a society has often been considered a measure of the economic well being of its people. But is it?

According to thermodynamics, the number of links in a human food chain is more a consequence of population density than of economic well being. As a rule of thumb, there is a 10% transfer of energy from one level of the human food chain to the next. Thus in China, where the people play the role of primary consumers, there is roughly a 10% flow of energy from the main agricultural crop (rice) to the human population (Figure 8.7). If another link was added to the food chain, so that the rice was fed to farm animals, and the people functioned as secondary consumers, the flow of energy would be about 1% (10% flow to the farm animals multiplied by a 10% flow from the animals to the people). By simple deduction, then, we can calculate that a two link food chain would reduce energy flow to the human population such that much of the Chinese population would starve. Even if the Chinese were rich, they would not be able to eat a lot of meat. Money cannot buy an exemption to natural law.

A.

B.

Figure 8.7 The Laws of Thermodynamics have a direct influence on human populations. (A) The one-link food chain of China. Although only 10% of the biomass of agriculture becomes human biomass, the energy transformation is sufficient to sustain a large population. (B) If the food chain was expanded to two links, the transformation of energy from agriculture to humans would be 1%, sustaining a much smaller population.

Having traversed much of The Living Staircase, we have gradually worked our way to the top—the biosphere. Here we can find another clear implication of entropy and thermodynamics. Imagine that you are an astronaut zooming into space in the year 3000 to explore other solar systems. Looking out the rear of the spacecraft, you see our small planet receding from view. It is a single yet complex entity that conforms to thermodynamics just as religiously as a cell. Energy enters the biosphere as light from the sun, and then is transferred and transformed in a million ways, losing heat at each step. For the biosphere to keep ticking, just as for a cell to keep metabolizing, the input of energy must continue. Some day in the future, the sun will deplete its potential for fusion reactions and stop emitting light—that will be the day when life on earth begins to run down, and shortly thereafter, flickers one last time like an exhausted candle.

Happily, that day won't come for another 5 or 6 billion years, and by that time, we may have moved to a new solar system, one with younger stars. This is the heartening spirit of the Case Western roommates on a cosmic scale: the Laws of Thermodynamics are inescapable. The room is going to get junky, and the biosphere will eventually run down. But what the heck? We can always move out![12]

? Q & A with Jeff & Jen

JEFF: You realize, Jen, that you've given college students the ultimate excuse to keep a messy room.

JEN: What do you mean?

JEFF: Well, entropy is an extension of natural law, and we all want to be law abiding, right?

JEN: Oh. Well, I suppose if it works, it works. That reminds me of the Student Government Association at Iowa State. When I was there, they passed a rule to outlaw the Second Law of Thermodynamics, saying that no reasonable legislative body would ever have enacted such a thing!

JEFF: I like it—clever humor is healthy. But let's move on to a serious question. The analogy between a dorm room and a cell has me confused about something. To clean up a room like the one you mentioned, you have to pile up a lot of garbage and throw it out. Cells don't do that, do they?

JEN: On a limited scale, they do. Some cells, especially the single-celled organisms, collect waste products in a vesicle and dump it periodically. But most cells recycle their waste products so efficiently that they don't accumulate. It would be similar to having a recycling plant in your own room, so you didn't have to throw things out.

JEFF: I see. So the analogy isn't quite perfect.

---

12. In fairness to Bross & Co., they didn't abandon the mess. By winning the national Pigsty Search, they won the benefit of a professional cleaning, in addition to a $1000 prize and a party for themselves and 100 friends. I hope that the Milton Bradley Company will still be around to help us out when the sun dies!

JEN: No, but your question brings up an important point. Living systems are extraordinarily efficient at recycling things—Chapter 3 gives a lot of examples—but they don't recycle energy.

JEFF: Why not?

JEN: It's a constraint imposed by the Laws of Thermodynamics. Any energy that enters will ultimately end up being lost as heat. If this didn't occur, a cell could recycle its energy, and it would become fantastically efficient.

JEFF: And then, they wouldn't have to have a constant input of energy to stay alive?

JEN: Right. A cell could take in energy for a period of time, cut off the supply, and recycle what it has. But according to Thermodynamics, it can never happen.

JEFF: Let's move on to my next question. Somewhere in the chapter you talk about producers, the components of a community that provide energy for all the rest, and you say that for most communities, these are plants.

JEN: Right.

JEFF: But why do you say "most"? What's the exception?

JEN: You're going to think I'm making this up, Jeff, but I swear it's true. Back in the 1970s, oceanographers discovered a community of organisms living at a depth of 2500 m. That's in total darkness, so plants are absent. As it turns out, there are a number of similar communities that live where the ocean floor splits open and lava flows out.

JEFF: I've heard of stranger things, I guess. So what are the producers there?

JEN: Bacteria that live off hydrogen sulfide, which forms when hot lava hits cool water. They use the energy of hydrogen sulfide in the same way that plants use the energy of sunlight—to make ATP and other organic molecules. All the rest of the organisms can trace their energy supply from there. One of the strangest things, though, is the animals that can be found. There are tube worms, for example, that are huge—a meter long! Everywhere else on earth, tube worms are a few millimeters in length. Other critters down there have never been seen before.

JEFF: How did they get there in the first place?

JEN: Good question. Nobody really knows for sure, but it's a topic that's being investigated. In 1991, an underwater volcano created a new hydrothermal vent in the Pacific Ocean, and it's being studied now. In just five years, it has been colonized by several species.

JEFF: This gives me a little hope about the future. If light isn't absolutely necessary for life, maybe we can hang around after the sun burns out.

JEN: Those bacteria only thrive in locations that are extremely hot. When the sun burns out, the heat of the earth will eventually dwindle. So your idea is theoretically possible, but realistically just a dream. Who would want to live on a world without light, anyway?

JEFF: Speaking of dreaming, what about your idea that we can all pack up and move to another solar system when the sun burns out?

JEN: Okay, I have to confess that I'm more optimistic than most. But if you consider that most of our scientific knowledge comes from the past 1000 years, and we have a good five billion years to go, it doesn't seem that extreme.

JEFF: I hope you're right!

RELATED RESOURCES

Bioenergetics and the determination of home range size, B. McNab, *American Naturalist* 97, 133–140, 1963.

Rebirth of a deep-sea vent, R. Lutz and R. Haymon, *National Geographic* 186, (November), 114–126, 1994.

Mitochondria—eclectic organelles, R. Lewis, *Biology Digest* 16 (October), 11–17, 1989.

Population, poverty, and the local environment, P. Dasgupta, *Scientific American* 272 (February), 40–45, 1995.

The Different Forms of Flowers on Plants of the Same Species, C. Darwin, University of Chicago Press, Chicago, 1986.

The Natural History of Pollination, M. Proctor, P. Yeo, and A. Lack, Timber Press, Portland, Oregon, 1996.

Time's Arrow: the Origins of Thermodynamic Behavior, M. Mackey, Springer-Verlag, New York, 1992.

Evolution, Thermodynamics, and Information: Extending the Darwinian Program, Oxford University Press, New York, 1987.

chapter 9

SEX

Theme

All organisms send and receive signals. Such interaction occurs among members of the same species and between organisms of different species. Communication between any two organisms is reflected in and is dependent on communication within each of their organ systems, organs, and the cells that compose those organs.

Examples

- Fertilization
- Receptors
- Neurotransmission
- Neuron development
- Intercellular communication
- Hormones and sexual maturation
- Courtship and mating
- Pheromones
- Territoriality
- Aggressive behavior

Three Months Ago

"**J**en, we've gone through practically an entire book and all we've gotten out of it is throwing in our two cents at the end of each chapter," Jeff mused. "What I want is a whole chapter to ourselves."

"It's a male thing, Jeff—control, what you want is control. But, seriously, I know what you mean. Actually I agree. I've been thinking about communication a lot lately. We haven't really discussed it specifically so far and it's an obvious central theme of life."

"No kidding, communication? Jen, we've been working together too long. My mind's been on the same subject. I guess because it keeps popping up when we've been discussing other topics on The Staircase, like the way FloJo's molecules interact or that nut Irv and his Division of Labor; in order for all the jobs in a cell or organism or community to be divided up, communication must be involved."

"Yeah, and remember our whole thing in the first chapter about emergent properties? Communication's thick with them. Communication is an emergent property of neurons, since communication among the neurons themselves is necessary for animals to communicate. And . . ."

"Slow down, Jen. I have an idea. Let's make some phone calls to the colleges around here and see if we can't get a hold of some scientists whose combined knowledge of communication will cover the whole Living Staircase. We'll invite them to a kind of mini-symposium. Communication is central to the work of anthropologists, psychologists, neurobiologists, entomologists, but I get the feeling the specialists from these areas rarely communicate with each other! I had a class with a biochemist who does research in fertilization. Maybe she can get us started."

"Organize it, and they will come."

Although it was much more difficult than they expected, after many phone calls and much rearranging (little if any person-to-person communication actually occurred throughout the whole process; a machine was almost always the intermediary), Jeff and Jen managed to set up a meeting of four local scientists who studied aspects of communication that covered much of The Living Staircase.

The Get-Together

"Welcome, everyone. My name's Jen, and this is my friend, Jeff. We'll be the moderators for this afternoon's discussion, which we've entitled 'Communication and

Sex: Molecules to Organisms.' Before I introduce our honored guests, I thought it would be valuable to briefly discuss the genesis of this meeting.

"After noticing that Communication was often referred to in passing over and over in previous chapters, Jeff and I wanted to explore the topic and how it fits into The Living Staircase. We needed a theme to focus our discussion on. Not surprisingly, Jeff suggested sex.

"After considering, I realized it really wasn't that bad of an idea. We put the concepts of communication and sex together in what we call the cycle of sex and reproduction. Yes, cycling again—just like in Chapter 3. You can see the cycle outlined here in this chart (Figure 9.1).

"Most animals, and even plants to some degree, move through this cycle. Opposite sexes attract each other, court, mate, and then their sperm and egg fuse to form an organism that develops into an adult of a particular sex. This organism then finds a new mate, and the cycle continues. We will refer to this cycle throughout our discussion. But, enough talking from me. I'll let Jeff introduce the folks who will guide us through this cycle and The Staircase."

"Thanks, Jen, and again welcome to everyone. Our guests today are, starting from your left, Dr. Doris Farthing, a fertilization biochemist and my former Biochemistry teacher (she passed me). Next to her is Dr. Martin Khan, a neurobiologist, who is especially interested in how neurons form their first connections in a developing embryo. Then we have Dr. Patrice Ning-Jones. She will enlighten us about hormones and their role in the cycle of sex and reproduction. Finally, Dr. Kisha Dryer is our expert on pheromones and courtship behaviors.

"Dr. Farthing, it seems reasonable to begin our cycle at the level of fertilization, since none of us would be here without it. Could you begin our trek by filling us in on how communication plays a role in sperm meeting egg?"

"Sure, Jeff, and, by the way, thank you for having us. I have a feeling I'll be learning a lot here today, also. We scientists tend to get tied up in our small corner of research. It'll be good to hear how what I do with cells and molecules relates to what the other folks here are studying.

"Since I'm first, let me just say that all communication, at its most basic, involves a sender, a signal, a receiver, and a response. In my particular area, **fertilization**, the egg is the signal sender, the sperm the receiver, and the response really involves both gametes (the general name for sex cells).

"I like to think of the joining of the two types of cells I study, sperm and egg, as a mini-version of two whole organisms getting together. The two gametes don't really court each other like the animals do in Dr. Dryer's studies, but, as you'll see, there's

Figure 9.1 The cycle of sex communication. Organisms of species that have distinct sexes court and then mate, producing offspring. These offspring develop and become sexually mature adults. Mature organisms of the opposite sex are attracted to each other, court, and the cycle continues. Each of us, and all organisms, are currently in the midst of this cycle. Communication is essential at each step. We will use examples from the cycle to illustrate how communication operates up and down The Living Staircase, from molecules to populations.

a lot of similarities in the two processes. Analogous to organisms about to court, the gametes first send out signals to attract the other cell, and then the cells 'mate'—that is, get together and fuse. The cell-to-cell communication of egg and sperm really relies on molecule-to-molecule communication."

"Excuse me, Dr. Farthing," said Jen, "but what do you mean 'cell-to-cell relies on molecule-to-molecule'?"

"Well, I think every level of The Living Staircase relies on every other. You know, each step holds up the others. But let me be more specific. The two gametes must attract each other somehow. This is especially dramatic in the case of organisms with external fertilization, like sea urchins and some seaweeds. They release their gametes out into the ocean! How in the world are they able to find each other?"

"You're right," Dr. Ning-Jones interrupted, "this *does* remind me of two organisms trying to find each other. If it's like what I study at the organismal level, pheromones, then the gametes locate one another by sending out some kind of chemical signal. The molecules you referred to, I suppose?"

"Exactly, Dr. Ning-Jones," Dr. Farthing went on, "for example, the sea urchin egg coat contains strings of ten amino acids (mini-proteins or peptides), called speract.[1] I can take some of this speract in my lab and add it to a tube of peaceful urchin sperm. As soon as I add speract, the sperm start swimming around like mad. The sperm must have receptors on their cell surface with which the speract communicates, binding to them and saying 'Follow me to the promised land, the egg.'"[2]

"Intriguing," interjected Dr. Khan, the neurobiologist, "this is nearly identical to the kind of communication that the type of cells I study, young neurons, go through in order to make their proper connections during embryonic development. Does a similar kind of communication go on in other animals?"

"Not only in other animals," the fertilization scientist continued, "but also in plants like the seaweeds I mentioned.

"Let me start with humans, though. In the lab, we can remove fluid from women's ovarian follicle cells surrounding developing, unfertilized eggs, and add this fluid to human sperm. These sperm behave just like the urchin sperm do when speract is added—that is, as if they are attracted to something in the fluid. The female seaweed gametes also secrete a special chemical that attracts male gametes. Strikingly, all of these 'courting molecules' are specific to each species. Even male gametes from one species of seaweed will not react to signals from female gametes of a different seaweed species. In fact, this is part of what makes species distinct."

"So," Jeff asked, "once they signal each other, how do the sperm and egg actually fuse?"

---

1. Another such peptide, AVP which controls monogamy in voles, is discussed in Chapter 7.
2. See Chapter 6 for a discussion of one type of "intermediary," called GTP-binding proteins, that is involved in getting signals from outside the cell to inside.

"Well, the two gametes have complementary pairs of proteins on their cell membranes. These proteins are similar to the receptors I mentioned and that I'm sure we'll hear more about today."

"Excuse me, Dr. Farthing, before you go on, could you give us a good definition for receptors, please, since it appears they'll be key to our understanding of communication?" Jen inquired.

"Sure, Jen. All the molecular receptors we'll be discussing are proteins, and they all have three basic parts, called protein domains as you can see in Figure 6.2; one that extends outside the cell (this domain often has different sugars attached that help in the specificity of binding), one that crosses the cell membrane, and one that extends into the cytoplasm of the cell. When something binds the portion of the receptor outside the cell, it changes the shape of the entire protein. This shift in shape is the signal that the cytoplasmic part of the protein uses to pass the word to the rest of the cell that something has bound the receptor. If you imagine yourself as a receptor, then it's sort of like if you'd tripped on a rock—only your lower domain interacts with the rock, but all of you, and also your shape, changes.

"So, sperm and eggs use these receptor-like proteins to bind each other much like a plug in a socket. This molecular communication sets off a sequence of electrical events. The series of *intra*cellular communication events in the egg resulting from the *extra*cellular signals of the sperm-egg interaction eventually leads to the sperm and egg membranes fusing, fertilization, and the beginning of the development of a new organism."

Similarities between Fertilization and Developing Nerves

"Dr. Khan, you mentioned how similar nerve cell communication was to sperm-egg communication. Would you care to elaborate?" asked Jen.

"Of course, Jennifer. We are now considering the organism after fertilization. Say, in the early embryos of mammals. We've moved on in the cycle of sex and reproduction far beyond our original two cells (but, remember, we used no new genetic information other than that with which we started!) to a complex, developing organism and its nervous system. We all know the nervous system is the brain's link to all the other systems of the body. In response to external stimuli, it sends signals to the other systems, like the muscles, so the organism can respond. Perhaps later on Dr. Ning-Jones will tell us about how different organ systems communicate in the cycle of sex.

"Anyway, the question I attempt to answer," Dr. Khan continued, "is: How do the millions of neurons that translate the stimuli get properly wired in the first place?" Dr. Khan, appearing pretty wired himself, was standing up now and gesticulating wildly in front of the group.

"I study this problem in grasshoppers, because they have large neurons we can easily observe under a microscope, and because they're relatively inexpensive to maintain."

"Excuse me," said Jeff, "but could we review the basic facts about neurons before we move on? It's important that we are all on the same wavelength in *our* communication."

"Definitely," answered Dr. Khan, hardly taking a breath. "Look at Figure 9.2 with me. Neurons are the cells that compose our nerves; they have long extensions called axons that are the neuron's 'arms.' **Neurons pass a signal** down their length electrically; and at the end of an axon, this electrical excitation is translated into a chemical signal, which then either passes on the signal to another neuron or to a cell of, say, the muscle system to excite or relax it. The chemical signal is composed of molecules called neurotransmitters, literally 'nerve senders.' They send the signal across a gap between one nerve cell and another called a synapse." Dr. Khan was on a roll—literally; when he shifted positions a crushed bagel was visible under his foot.

"Remember how speract sends a signal to the sperm by interacting with a receptor on the sperm surface, which then signals the sperm to move towards the source of the speract? Well, in a similar way the neurotransmitter interacts with a receptor on the other side of the synapse, signaling the receiving cell to pass on the message. Make sense?"

"Yeah," said Dr. Ning-Jones, "but I thought you were going to tell us how neurons got into the right position in a developing organism. I mean aren't we trying to get our organism developed and wired correctly, so that it can eventually send and receive all the right signals, so that it can mate and make new little developments of its own?"

"Good point," Dr. Dryer chimed in, a little harsher than she probably meant to, "All this talk about *molecules* and *cells*. How about the organism itself? Where does it all connect?"

"Patience, ladies, patience. A developing grasshopper's neurons grow down specific tracks and contact specific sets of other neurons seemingly in a very repeatable, programmed way, as you can see in this Figure 9.3. The axons on these tracks arrange themselves in ladder-like bundles."

"But, Dr. Khan, how do they know which bundle to get into?" Jeff asked, clearly intrigued.

"This is where it gets wild. Each neuron literally explores its environment, 'sensing' the right way to go! The exploring end is called a growth cone, and it has tentacles called filopodia that radiate from it and communicate with a jungle of neighboring growth cones. Sometimes I picture the developing nervous system like

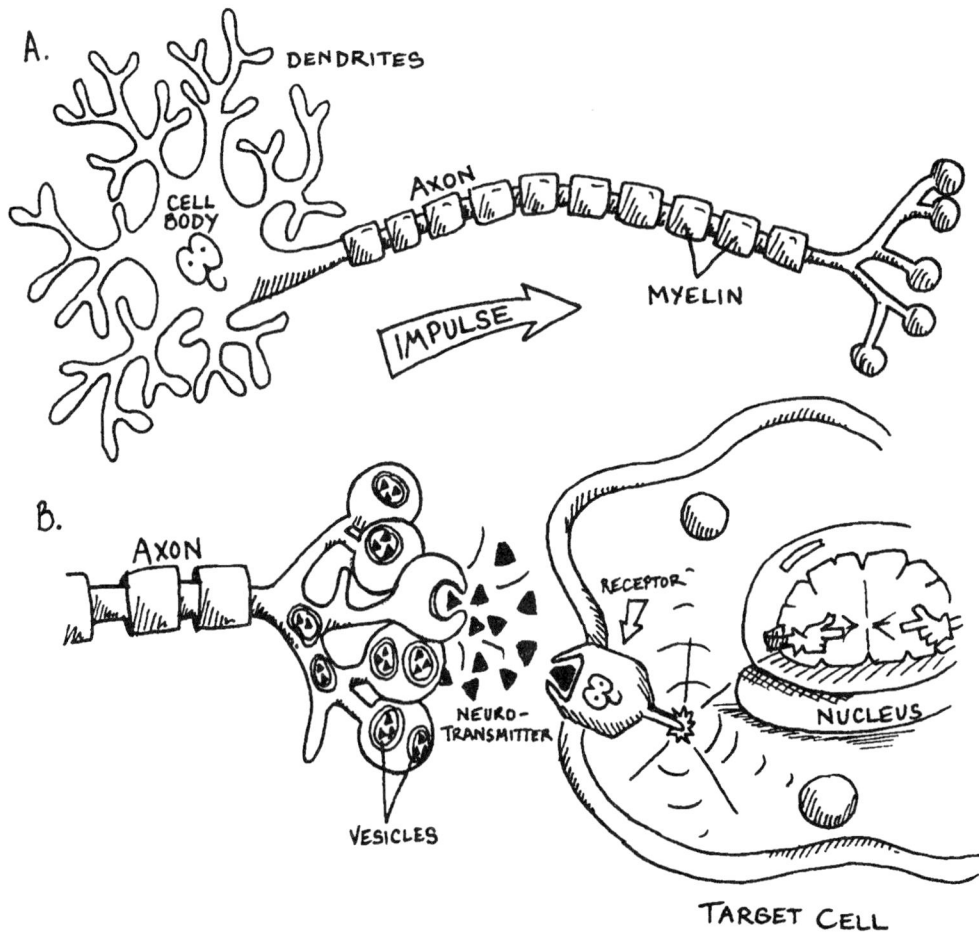

Figure 9.2 Neurons. (A) A neuron is a complex single cell composed of a cell body that sends signals via extensions called axons and dendrites. During development, neurons grow toward each other and establish connections. (B) The end of an axon communicating with a target cell, which could be a muscle cell or another axon. An electrical nerve impulse travels down the axon to the end, where it excites the vesicles which store neurotransmitter to fuse with the plasma membrane of the neuron and release their contents into the space between the releasing neuron and the next cell. This chemical message then binds receptors on the receiving cell and the message is propagated.

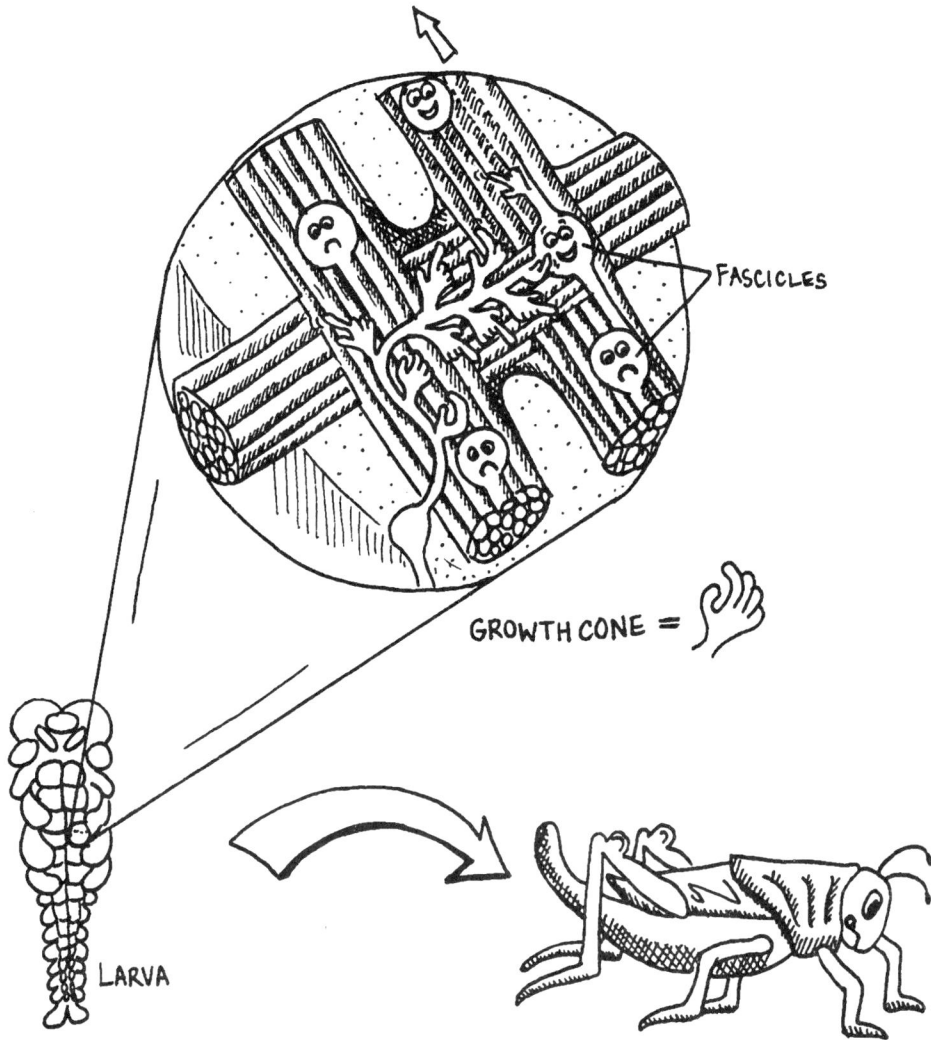

Figure 9.3 Neural development is relatively easy to study in grasshoppers because of the large size of the organism and the ordered growth of the neurons. The axons grow down tracks of axons that were previously laid down. The growing ends of neurons, called growth cones, move, turn, fuse, and stop depending on signals they receive from surrounding cells and their eventual targets. A fascicle is a bundle of closely associated axons.

thousands of people all released into the same arena at the same time, each with directions on where to go, how to get there, and somebody at their eventual destination calling out for them. Sounds chaotic, I know. The things doing the 'calling' in this case are what struck me about the parallels between Dr. Farthing's sperm-finding-egg story and my neuron-finding-correct-connection story. The neuron's eventual point of connection sends out a signal to call the neuron, just like speract calls the sperm. A receptor on the neuron's surface senses the signal and follows it!" Dr. Farthing perked up, emphasizing her point by slamming her hands vehemently on the table. Everyone jumped.

"Exactly, Dr. Farthing," Dr. Kahn said recovering, "the filopodia feel around, contacting different cells and if they don't stick well to a surface, they retract. If they do adhere well, they create a tension that guides the growing cone toward its eventual site of attachment. Adhesion is the result of molecules on the searching filopodia 'matching up' with recognition molecules on the receiving neuron surface. Again, this surface-to-surface interaction is very similar to the interaction between gametes, but it gets even better.

"Look at this picture (Figure 9.3). Here, different growth cones follow the same axon bundle up to the end of a particular horizontal bundle. These growing neurons, following a pause, send out explorers that lead to their growth in three different directions. Later in development, several more growth cones will emerge from these axons.

"Each time a filopodium finds its partner neuron, by matching up cell surface receptors, it inserts into the growth cone of that neuron. Search, recognition, and fusion! Sound familiar!?

"We think that insertion of filopodia into growth cones may induce new types of recognition receptors to be made on both the inserting and inserted-into neurons. This would result in the neurons having entirely different recognition surfaces than they had prior to meeting. So now, new and different cones will recognize these neurons.

"Insertion of filopodia into cones," continued Dr. Khan, running out of steam, "might also signal neurons to make new growth cones. These cones also would grow, and the finely-tuned construction of the grasshopper nervous system would continue building on itself. In this way, we believe, a complex nervous system arises; after searching, finding, adhering, and fusing, neurons recognize and grow new neurons and, in turn, new neurons recognize them. Here, at a cellular level, neurons are both senders and receivers of signals. A beautiful example of communication, don't you think?"

Jen took a deep breath and said, "Thank you, Dr. Khan. Let's take a short break before continuing around the cycle and up The Staircase, shall we?"

At the break, Jeff noticed Drs. Khan and Farthing in close communication themselves. Although he tried not to look, Jeff couldn't help thinking that, apparently, the good doctors had been looking at more than just Figures 9.1–9.3.

Inside Communication

"Well, folks," started Jeff, "I think we're getting a perspective on communication that few of us could have appreciated before today.

"Before we leave this part of The Staircase altogether, I was wondering if we might take a step back down and into the cell for a minute. I was struck by another parallel between Dr. Farthing's story and yours, Dr. Khan (Jeff tried hard to look at neither of them). You both talked about things happening outside the cell that caused changes inside the cell. But neither of you got very specific. Can you, Dr. Khan?"

"Sure, Jeff. **Intercellular communication** is, of course, enormously important. We do study some of those events in my lab. For example, organelles, the membrane-bound subcompartments within cells, also communicate with each other. Proteins and other molecules are shuttled precisely from organelle to organelle and from inside the cell to outside or vice versa. Each organelle is bound by a fatty membrane much like the one that surrounds the cell in which they reside."

"Dr. Khan, if you'll excuse me, you're sounding like a textbook," piped up Dr. Ning-Jones, who, since we began this session, had apparently been dozing. "Get to the point. What does this have to do with Dr. Farthing's communication story or anything for that matter?"

A little taken aback, Dr. Khan cleared his throat and continued: "Uh, yes, like I was saying, I also work on intercellular communication. Since we're talking about neurons and how they develop in grasshoppers, I'll describe some of what we study in terms of the nervous system.

"Remember how we discussed the neuronal signal changing from electrical to chemical and then back to electrical?[3] Well, the chemical part of this signal transmission involves neurotransmitters that the cells store in membrane-bound bags, or vesicles, in the tips of neurons (Figure 9.2). When the neurons are stimulated, the bags move to the extreme ends of the neurons, where they dump their neurotransmitter into a small space, the synapse we heard about a few minutes ago, between neighboring neurons, and then pass on the message of stimulation to the next neuron, which converts the signal back to an electrical one and so on."

---

3. Transformations of this type are also discussed in Chapter 8, where the emphasis is on energy exchange.

Dr. Khan looked at Dr. Ning-Jones out of the corner of his eye to make sure she was getting the message. "One neuron is the sender, another the receiver. Neurotransmitter is the signal. How does all this intercellular communication happen? Who carries the mail (the proteins and other molecules) from compartment to compartment around the cell? Who are the mailmen?"

"Dr. Khan," Jeff said politely, "I'm lost. Where did the mailmen come in?"

"Just an analogy, Jeff, the mailmen are the vesicles that carry and deliver the mail, you know, the important signals, inside the cell. Yeast cells and human brain cells, and all cells in between, have similar mailmen or vesicles. These vesicles are organelles unto themselves, contained by their own membranes."

Dr. Khan stopped for air as if to see if there were any questions. Jeff could've sworn he saw a discreet exchange of glances between Dr. Khan and Dr. Farthing, or was he imagining?

When there were no questions, Dr. Khan went on, "A classical mail route is the one followed by secreted proteins. Like all proteins, those which will be secreted are synthesized in the cytoplasm at the ribosomes (Figure 9.4). Proteins that are destined to be secreted from the cell contain a signal that sends them from the ribosomes into another organelle, the endoplasmic reticulum (ER). The ribosomes where these proteins are synthesized actually and conveniently sit right on the ER. Once in the ER, the proteins are packaged in a vesicle that buds off the ER. This first mailman is a vesicle that brings the protein to the Golgi apparatus and stops and docks, membrane to membrane.

"This is why I got all excited when Dr. Farthing was talking about how sperm and egg fuse. Very similar communication occurs between the vesicle membrane and the membrane of its receiving organelle. In this case the vesicle carries the protein from the ER to the Golgi. The vesicle and Golgi membranes fuse and the protein is delivered into the Golgi. Then another vesicle carrying an altered version of the same protein buds off from the Golgi and moves to the plasma membrane, where the same thing happens again! This time the protein winds up outside the cell."

The audience listened, rapt. Dr. Khan was a communicator par excellence. "Remember, this mail service is going on in every kind of cell involved in every kind of process: including the pheromone sending and receiving and the courtship and mating behaviors we'll hear about shortly. Each cell specializes in particular mail routes. In speract-secreting cells (eggs), the main mail is speract; in nerve cells, neurotransmitters make up the bulk of the mail. The general mail routing plan, however, is very similar in all cells."

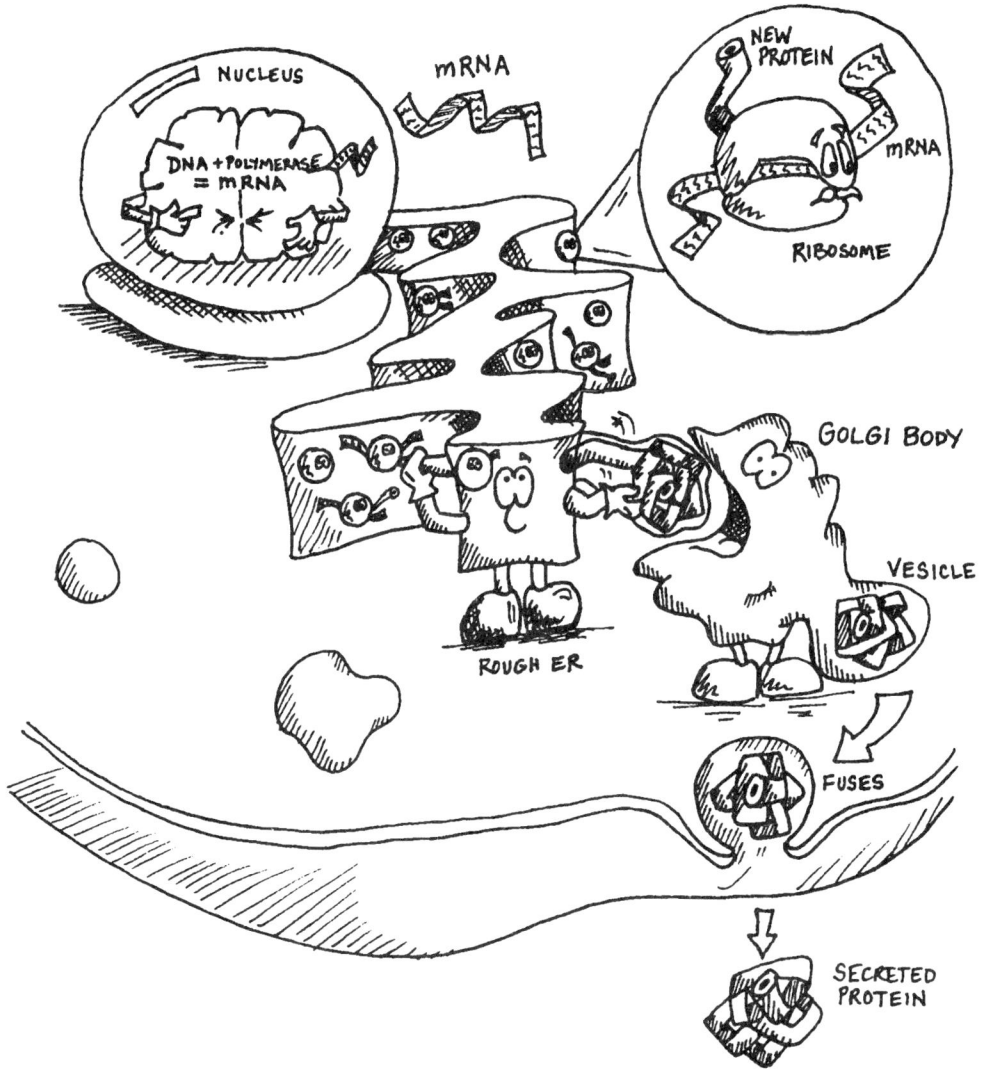

Figure 9.4 Schematic of the voyage of a secreted protein. (1) DNA is made into mRNA (see Chapter 4 for details) that leaves the nucleus (in eukaryotes) and (2) is translated at the ribosomes residing on the endoplasmic reticulum (ER). The protein enters the ER for modification and then (3) leaves the ER in a vesicle that (4) moves to the Golgi for further preparation for secretion. Finally, the protein leaves the Golgi in another vesicle, (5) fuses with the plasma membrane and is secreted from the cell.

Hormones, Puberty, Organ to Organ

"Thanks again, Dr. Khan," Jeff began, "Dr. Ning-Jones, I believe it's your turn. When we move around the cycle of sex and reproduction and along The Staircase, we come to the maturing organism that needs to mature some more, if you know what I mean. We need to discuss communication at the organ system level to understand how organisms sexually mature. This is where your research fits in, no?"

"Glad to make a contribution, Jeff. Systems communication, that's where me and my hormones come in . . . I mean not *my* hormones, specifically." Dr. Ning-Jones blushed, pretended she hadn't, and quickly continued: "Hormones communicate inside organisms, between the nervous system Dr. Khan discussed and the reproductive and other systems. The hormones and the glands that secrete them compose what we call the endocrine system. Hormones that come to mind in our context of sex and reproduction are the ones you suggested, Jeff, those responsible for **sexual maturation**, or puberty, as we call it in humans. Sexual maturation must occur before organisms can have any offspring."

"Geez, I hated puberty," said Dr. Dryer a little too loudly. Everyone looked at her, having forgotten she was there. She smiled and the whole room erupted in laughter.

Dr. Khan brought us back to the subject, "Pardon me, if you would Dr. Ning-Jones, but since I guess hormones are the signal and receptors receive them, who sends them? Where do they come from?"

"Good question, Dr. Khan. And Dr. Dryer, we all empathize with you," Dr. Ning-Jones continued smoothly. "In humans, somewhere between the ages of eight and fourteen, it is the hypothalamus on the ventral part of our brains that sends the signals. For sex maturation, we call the hormone-signals gonadotropins.

"As you can surmise from their name," Dr. Ning-Jones continued, "gonadotropins affect the gonads or sex organs. As you mentioned, Dr. Khan, and keeping with the theme we've heard again and again, these hormones bind to receptors on specific gonad cells and induce the cells to secrete other sex hormones, such as androgens and estrogens."

"Ah," said Jen, "we heard about those guys in the context of regulation before. Something about using artificial androgens to increase athletic ability.[4] Sorry, Dr. Ning-Jones, please go on. What do these hormones do in their normal context?"

"In women, estrogens initiate and control the ovulation cycle, activate development of the breasts and growth of pubic and underarm hair, increase water retention, and affect sexual behavior. Androgens are responsible for all male sex characteristics:

---

4. In Chapter 4, there is a discussion of the regulation of the sexual maturation process.

testes development, sperm production, muscle growth, deepening of the voice, and growth of facial, pubic, and underarm hair."

"Y'know, Dr. Ning-Jones, hormones also signal egg maturation prior to fertilization," interjected Dr. Farthing.

"You are quite correct, Dr. Farthing," responded Dr. Ning-Jones, "in fact, even long after fertilization, hormonal communication continues across the placental bridge between mother and developing embryo. The embryo sends messages to the mother's reproductive system, such as those telling it to stop her menstrual cycle. If the cycle continued, the result would be a miscarriage. Signals from the embryo also induce the formation of a protective plug in the mother's cervix and increase the size of her uterus and breasts."

"That reminds me," Dr. Dryer interrupted, perhaps trying to redeem a bit of credibility after the puberty remark, "In my field of sexual behavior, I was reading an amazing account of some research that demonstrated how a single hormone can control a complex mammalian behavior which is part of the sex and reproduction cycle. Scientists studying the prairie vole have recently shown that one particular hormone causes that animal to be monogamous, that is, to mate with only one other vole, as opposed to having a variety of mates.[5] Imagine, one hormone responsible for one mate. It seems hormones play a role in controlling many parts of the sex and reproductive cycle."

To himself Jeff wondered if Drs. Khan and Farthing were thinking about this very cycle in relation to each other, and he started to feel sort of embarrassed. Trying hard to avoid eye contact with them, he said, "Dr. Dryer, that's a nice segue into looking at your work on pheromones and courtship and how organisms interact in the next part of the sex and reproduction cycle. In fact, aren't pheromones much like that vole hormone you're talking about?"

Courtship and Pheromones: Spiders, Butterflies, Yeast and Us

"They are and they aren't," Dr. Dryer answered. Pheromones are like hormones in that they are signals that influence behavior and physiology, but, and it's a big but, pheromones are emitted *externally*, and therefore are sent organism-to-organism, instead of organ-to-organ within an organism.

"I'll concentrate on **pheromonal influence on courtship and mating behavior** here, but animals also use distinct pheromones to identify different members of a community, to serve as navigational signals, for mother-infant bonding, to deceive predators or prey, or to regulate development. As we've seen with all communication,

---

5. See discussion of this hormone, vasopressin, and the monogamous vole in Chapter 7.

pheromones and the behavior they influence involve a sender, a signal, a receiver, and a response."

"How about humans, Dr. Dryer," said Jen with a smile, "I mean for example, might some of us in this room be pheromoning right now?"

Jeff looked at Jen with disbelief.

"Quite possibly," Dr. Dryer responded without missing a beat, but with a smile. "We even have suggestive evidence for the existence of plant pheromones."

"Say, Dr. Dryer," interrupted Dr. Khan, "my St. Bernard, Herman, has practically leapt his six foot fence to reach a female dog in heat. Do you think pheromones may actually control all male-female sexual interaction?"

"That's a loaded question, Dr. Khan, and I don't think we have enough information to answer it, nor do I think we can ever exclude the environment and other factors, like brains when it comes to us, for example, from having an important role in behavior. But pheromones are certainly key to courtship behaviors. Listen to the story of the danaid butterflies, and see what you think.

"It starts with a plant," continued Dr. Dryer, taking off her glasses and standing up as if to emphasize what she was saying, or perhaps, to take on the characteristics of the particular plant in question. Everyone was spellbound, receiving Dr. Dryer's signals with intensity.

"The plant contains toxic compounds that make it taste really bad to your average bug. The danaid butterfly, however, eats the plant with gusto. After the male butterfly obtains this chemical from the plant, he stores most of it in special glands on his rear-end called pencil hairs. He uses the rest of the compound as raw material for making a pheromone, which he stores on the pencil hairs to help attract females.

"When the male meets up with a female, he checks to see if she's receptive to him by rubbing his pencil hairs on her body. Some behavioral entomologists hypothesize that the female's response is based on the amount of pheromone she senses on the pencil hairs of the courting male. The more pheromone, the reasoning goes, the more alkaloid the male has eaten, the more he has stored in his glands, the more receptive the female."

"Why in the world, Dr. Dryer," burst out Dr. Ning-Jones, causing Jeff to start, "why in the world would a female want a male who has the most *poison*?!!"

"Well, Dr. Ning-Jones, as the male copulates with the female, he transfers the alkaloid to her, and she in turn transfers the chemical into her eggs. The poisonous chemical protects the eggs from predators."

"Amazing," several members of the panel mumbled together.

Hearing no questions, Dr. Dryer went on, "Another pheromone altogether, in male mouse urine, causes a new female mate to spontaneously abort embryos that he did not fertilize. Talk about communication!

"And finally, Dr. Khan, to get more directly to your original St. Bernard question about pheromones controlling courtship and mating. We have good evidence for such a pheromone-controlled scenario in millipedes. After the female millipede has attracted a mate, the male secretes an odorous pheromone, which the female eats. This pheromone causes her to freeze in a sexual presenting position until the pair finish copulating. And, in other insects, scientists have identified pheromones that cause males to pump up and down during copulation.

"We think that much like in the other types of communication we've heard about today, specific molecules and cells must be responsible for these behaviors. In the scenarios I've described, each male movement is the stimulus for a female movement and vice versa. If any part of the ritualized sequence breaks down, the pair starts the whole process over." Dr. Dryer sat down, looking tired.

Jen entered the conversation: "Dr. Dryer, in thinking of how to connect pheromones and their power, I mean, their influence, in connecting them to the rest of The Living Staircase . . . Just like for speract and the developing nerve and all communication, really, doesn't there have to be some receptors for the pheromones? Receptors on some kind of cell?"

"Good question, Jen," Dr. Dryer responded. "The answer is 'yes,' and the best research on such receptors has been done in yeast. Yeah, even unicellular organisms use pheromones. In baker's yeast, the very yeast that makes your bread rise,[6] mating pheromone attracts partners by binding to protein receptors on their cell surface. The receptors then pass on the signal into the cell, inducing a response. The response includes: (1) halting division of potential 'mates' (yeast can either divide on their own by mitosis or find a mate and undergo meiosis) and (2) activating specific genes in the receiving cell that encode proteins that cause the two yeast cells to align prior to mating.

"And," continued Dr. Dryer, "I've got to tell you about this incredible communication study on humans. It sort of relates back to Jen's question about human pheromones and it fits in well with your cycle of sex and reproduction, too, Jeff—right at the mating step.

"Many proteins are involved in helping the immune system recognize and destroy foreign molecules. Some of the most important are called MHC, for major histocom-

---

6. The usefulness of yeast for studies relevant to other organisms is also discussed in Chapter 6.

patibility complex. Well, it seems that the type of MHC you have influences both your body odor and your body odor preference!"

"What," said Dr. Ning-Jones, "are you talking about?"

"I'll back up," continued Dr. Dryer. "The thought is that one benefit of sexual reproduction is that it allows animals to react quickly to an environment of invading parasites that is constantly changing and often threatening. So, in the best case, females would be able to provide their offspring with combinations of genes that would efficiently arm them against such invaders. MHCs are molecules that can do this; so, generally speaking, the more different the MHC genes (there are at least five) in each parent, the more combinations of proteins their offspring can make, the more foreign molecules they recognize and then destroy. Make sense?"

"Sort of, but what does this have to do with communication?" queried Jeff.

"Hold on and you'll see. The experiment went like this. First, the scientists determined the MHC-type of several female and male students. Then, each male student wore a T-shirt for two consecutive nights. The next day, the female students rated the T-shirt smells. And it turned out the women preferred T-shirt odors from men with MHC-types that were the *most different* from their own. Also, these same smells reminded the women more of former mates than odors of men whose MHC-type was similar to their own. Can you see? This suggests that MHC-type is some kind of signal, received by the olfactory system, that influences mate choice!" Dr. Dryer finished triumphantly. Some people in the room appeared stunned; others shook their head, not believing what they'd heard.

"An engaging story, Dr. Dryer," Jeff said quickly, not wanting to think about Dr. Farthing smelling Dr. Khan's shirt, but doing so anyway. "Much food for thought. Before we finish up, let's take another short break, okay?"

A Movie and Territoriality and a Walk Down The Staircase

"Well, folks, we're nearing the end of our time together," Jen began. "I was wondering if any of you scientists had any questions for each other?"

"I do," Dr. Ning-Jones piped up. "We haven't really discussed communication at the level of populations, which we really ought to do."

"Funny that you mention that, Ning-Jones. While I was drinking my tea during the break, I thought of a great example of just such a type of communication," Dr. Farthing, who had been listening for the last hour, joined in. "And it's even related to the cycle of sex and reproduction," she began.

"Once animals find a mate, they often **establish a territory** using chemical signals in their urine. It's sort of a communication that tells other individuals of the same or different species to stay away.

"Anyway, in a movie, *Never Cry Wolf,* I saw the other night, this scientist is studying wolves and manages to find an Arctic wolf den. He sets up a campsite nearby. Curious, the male of the wolf family (these wolves mate for life) comes over to investigate the intruder and proceeds to mark off a territory with his urine near the scientist's camp and all around the wolf den.

"The scientist feels the wolf's territory claim leaves him with too little room to perform his study. So, after drinking dozens of cups of tea, he marches out and uses *his* urine to mark off a territory that gives him a bit more room. The wolf responds by again marking his territory, but this time just on the other side of every spot where the scientist had left his mark."

"I've had too much tea myself. Have to go mark some territory. Be right back," said Dr. Khan.

After everyone had drink refills, Dr. Khan returned and spoke up, "I like that story, Doris . . . uh, Dr. Farthing. I was thinking of something along similar lines of groups of organisms communicating. It seems to me a lot of **territorial and aggressive behavior** is connected to the cycle of sex and reproduction we've been discussing. Have you run across such behavior in your research, Dr. Dryer?"

"Hmm, and I wonder if there's such a thing as an 'aggression pheromone' that instead of saying 'come hither' says 'get away'?" Dr. Ning-Jones said mostly to herself.

"Probably is, Dr. Ning-Jones," answered Dr. Dryer. "I mean animals often perform ritualized dances of aggression, which are very similar to those courting dances I mentioned. Most act out an intimidation ritual without actual physical contact, which would risk injury. Male fiddler crabs, for example, raise their one huge claw threateningly, and jumping spiders brandish brightly colored hairy patches and alter the position of pigments in their eyes to make them look like they're flashing.

"Animals like baboons also use visual and physical communication to create and maintain their rank within a family group. Baboons will first raise their eyebrows and stare at an offending party. They'll follow that with a nip of the offender's neck, if he doesn't fall in line," Dr. Dryer continued.

Deception

"There's a nasty twist on communication in our cycle of sex and reproduction," said Dr. Dryer. "Even though we've probably all experienced it, we haven't talked at all about deceptive communication. And I know a case of it you won't believe."

Everyone listened intently as she went on, "It's about the female bolas spider. Remember pheromones? Well, male moths are lustfully attracted to the female's mating pheromone. The female bolas uses this to her advantage. She builds a very unusual web, which consists of one horizontal line and a vertical line hanging from it with a sticky ball of silk at the end.

"She positions herself so she can swing her web-pendulum and begins secreting imitation male moth-attracting pheromone that is very similar to the one the female moths secrete. When the unsuspecting moth zooms up expecting a receptive female moth, the bolas swings the pendulum. A couple of times a day, she manages to stick the ball to a moth. Then she climbs down her pendulum of death and eats him."

"What a way to go," said Dr. Khan, "expect sex, get death."

Summing Up

"Anyway," said Jen, "all communication sticks to the same general theme of sender, signal, receiver and response, but it seems like there's almost as many variations on this theme as there are different organisms. So many similarities, but so many differences.

"I was just mentioning to Jeff during the break that a useful way to sum up our discussions would be to follow one example of communication which spans the entire Living Staircase."

Dr. Dryer volunteered to walk The Staircase. "Okay, let's see," she gathered her thoughts, "Since I mentioned baboons, and I've recently been reading about them, let's take their sex cycle as our case study. The female baboon presents a mass of brightly colored tissue on her posterior when she is in heat. When a male baboon sees this display, he becomes stimulated and pursues the female. This is primarily a visual signal, but I'm sure pheromones are involved somehow, too.

"Look at this drawing I just happen to have here with me (Figure 9.5)," Dr. Dryer said dryly. "The light making up the female image shines on protein receptors in the baboon's eye. Light is the signal; the eye, or more specifically, cells in the eye, contain the receptors. The receptors, as we could guess, span the membrane of vision cells. They are the bridge from the outside of the cell to the inside. In basic structure and function, they are very similar to the other protein receptors we've discussed, like the ones for speract in urchin sperm, and the ones for pheromones in yeast.

"The receptors pass the information of the female-in-heat image to other proteins inside the cell, which in turn pass the information to other proteins. Some of these proteins are ion channels on the cell surface. Ions are charged atoms, and when ions shift across cell membranes through these channels they create an electrical energy

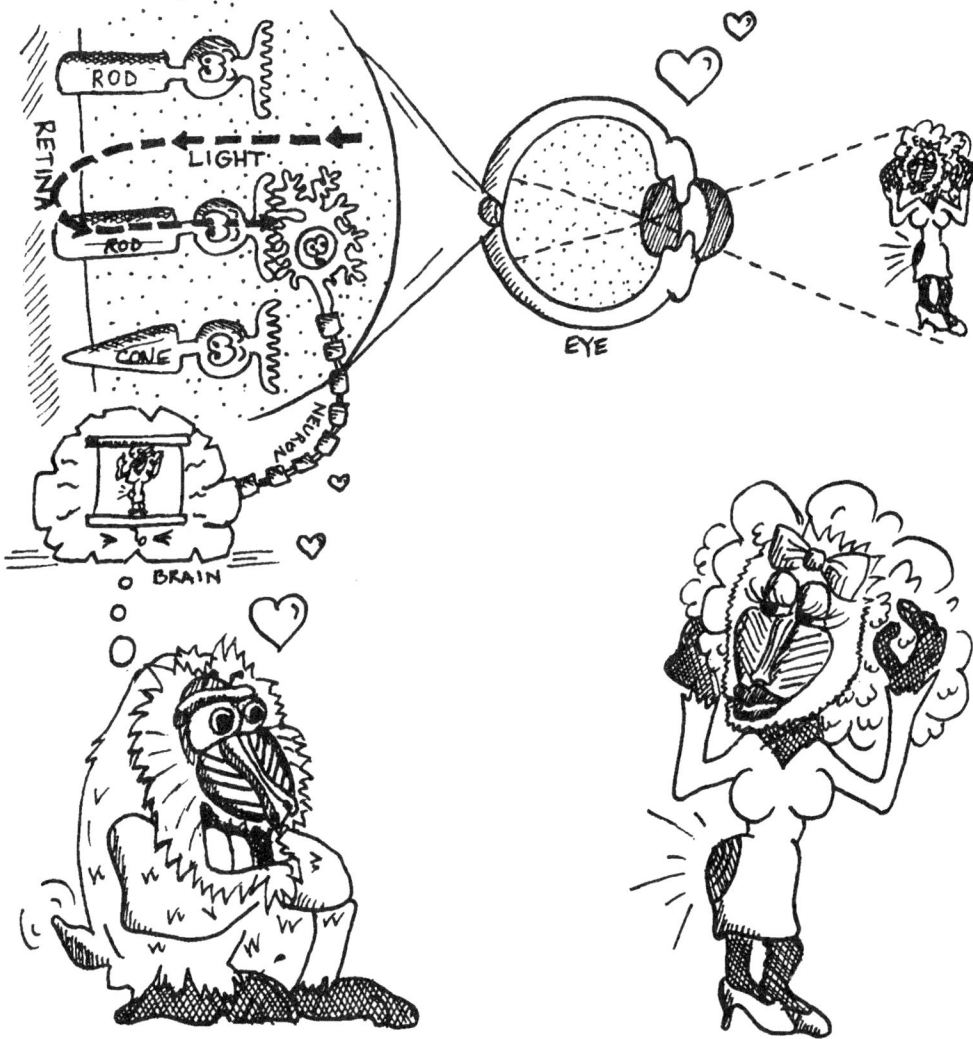

Figure 9.5 Visual communication. The image of a female baboon in the form of light energy shines on the retina of a potential mate. In the retina reside cells that contain photosensitive receptors. The best studied of these cells are the rods, which are responsible for monochrome vision in low light and for color vision in bright light. Rods are special cells which have an outer segment filled with photosensitive receptors and a portion that contacts a nerve cell to pass on a stimulatory signal. The light energy alters the receptor's shape, which leads to a change in the shape of ion receptors in the rod plasma membranes, blocking the flow of ions. Repeated many times over, this signals the rod's nerve cell portion to send the message of the female image. This signal travels the nervous system to excite the muscles of the baboon, and he responds to the female. Chapter 4 describes this process of muscular contraction necessary for his response.

that is passed from cell to cell, bridge to bridge, in the nervous system, as Dr. Khan talked about earlier.

"This electrical energy, still holding the image of the female, is translated to chemical energy and back to electrical again as it travels through the nervous system to the brain. The brain interprets this signal and then sends a signal back through other nerves to receptors (on, for example, leg muscle), stimulating them. Then the male baboon takes off after a potential mate. This whole process of conversion of light energy to chemical energy to electrical energy takes only milliseconds. There you have it—molecules, cells, organs, organ systems, organisms, all interacting in a population. And I should throw in that all five senses rely on very similar mechanisms of communication at the molecular level," Dr. Farthing finished up, out of breath.

"Bravo," shouted Dr. Khan, endearingly. "Let's all go get some dinner." And with that the meeting "Communication and Sex: Molecules to Organisms" came to a close.

Q & A with Jeff & Jen

JEN: Good meeting; I'm glad we finally got to do a chapter. But we hardly touched on the major type of communication that human beings use, that everyone at the conference was using: language. I've always wondered how we learn to speak, from "gaga" to full sentences within a matter of a couple of years. Don't you wonder, Jeff, how this magic happens?

JEFF: Actually, during the conference, I started to think about language a lot myself. I did a little bit of looking into how we learn language. I found out some fascinating stuff. For example, *at six months of age* we're already able to distinguish meaningful from meaningless sound variations, and only in our native language.

JEN: Who figured that out?

JEFF: Patricia Kuhl and her colleagues devised a test during which the infants sit in a parent's lap and then listen to repeated sounds from a speaker near them. The infants are trained to turn and look at the speaker when they hear a variation in the sounds, at which time they are rewarded by the appearance of an animated toy like a stuffed panda playing drums.

JEN: So, in these tests they showed that the infants would turn and look at the speaker more often when variations were made in sounds from their native language than when variations were made in sounds from a foreign language?

JEFF: Yep. They had a group of Swedish babies and one of American babies and used sounds unique to their respective native tongues. From her results, Dr. Kuhl concluded that by six months of age, we already are developing a "language filter," so that instead of recognizing all language sounds like we probably can at birth, we begin to focus only on those "relevant" to us.

JEN: That is cool. But how do the infants know what sounds are important and meaningful without even knowing the meaning of words?

JEFF: Scientists don't know yet, but one guess of Dr. Kuhl's is that it may have something to do with baby-talk, that is, she thinks that maybe the slow, silly way we talk to babies is involved in the process of babies differentiating which sounds have meaning.

JEN: Interesting. If we begin to filter sounds around six months of age, when and how do we start making some of these sounds back to the world?

JEFF: That's another active area of research, Jen. It's been known for some time that we start babbling between seven and ten months of age, but scientists didn't really begin to understand the mechanism of this babbling until recently.

JEN: Babbling has a mechanism? I mean isn't it just babbling?

JEFF: Not really, babbling is key to language development. Laura Petitto and Paula Marentette made some very intriguing observations about babbling when they studied babies who were born deaf and had deaf parents. These babies saw communication only by sign language. The researchers compared them to babies who could hear and had parents who communicated by speech. They found that, in addition to babbling verbally, the deaf babies actually babbled in sign language!

JEN: What? What do you mean?

JEFF: Well, the researchers analyzed the hand movements of the deaf infants compared to hearing ones and showed that deaf infants had a sign-language-based structure to their hand movements. They also found that this babbling, which they call manual babbling, has all the same characteristics as vocal babbling. It starts at the same age and has the same timing and stages of development as vocal babbling, including the stereotypical syllabic repetition, like the goo-goo, gaga, you were talking about.

JEN: That means that learning language doesn't rely on speech or hearing!

JEFF: Apparently. The scientists conclude humans have an inborn brain-based "language potential" that is linked to an "expression potential" capable of processing language and expressing it manually or vocally. The researchers think babbling is what bridges expression potential and actually talking.

JEN: This reminds me of the discussions of natural selection, homology, and diversity in Chapters 5–7, Jeff. It sounds like language development is like other systems in biology discussed in those chapters[7] in that it is both limited and flexible. Language development is constrained by structure. It can use only syllables or phonetics. But it is flexible in that the expression of this structure can change depending on the environment.

RELATED RESOURCES

How Embryonic Nerve Cells Recognize One Another, C. Goodman and M. Bastiani, *Scientific American*, Dec., 28–66, 1992.

Secrets of secretion revealed, M. Barinaga, *Science* 260, 487–489, 1993.

Animal Communication by Pheromones, H. H. Shorey, Academic Press, 1976.

MHC-dependent mate preferences in humans, C. Wedekind, T. Seebeck, F. Bettens, and A. J. Paepke. *Proceedings of the Royal Society of London Behavioural Biological Sciences* 260, 245–249, 1995.

Social Organization of Hamadryas Baboons, A Field Study, H. Kummer, the U. of Chicago Press, Chicago, 1968.

Priming the brain's language pump, M. Barinaga, *Science* 255, 535, 1992.

Linguistic experience alters phonetic perception in infants by 6 months of age, P. Kuhl, K. Williams, F. Lacerda, K. Stevens, and B. Lindblom, *Science* 255, 606–608, 1992.

Babbling in the manual mode: evidence for the ontogeny of language, L. A. Petitto and P. Marentette, *Science* 251, 1493–1496, 1991.

Chemical Communication, the Language of Pheromones, W. Agosta, *Scientific American Library*, 1992.

---

7. See, for example, the discussion of the female wasp *Nasonia vetripennis* and her egg-laying behaviors, which are limited by her physical structures, but vary depending on the environment in Chapter 7.

chapter 10

WALKING IT

Theme

After having pondered Living Staircase examples from the eight central themes of Biology, you should be able to fit any new biological information you learn into a broader context. In other words, you should be able to take new information and walk it to its logical location on The Staircase, fitting it snugly amongst the relevant bits of knowledge you already have. The central themes will serve as a handrail to guide you.

Examples

- Neurotransmission
- Cancer

We've been through quite a lot together. As we listened to the different lives of Ira Bloomfield, pondered the philosophy of Tibetan monks, sprinted through the Olympic trials with FloJo, and read the musings of Carl Linnaeus, we began to grasp great themes of biology: communication, division of labor, cycling, regulation, natural selection, homology, diversity, and energy flow. We saw how examples of each of these themes could be found throughout the entire Living Staircase, stretching from molecules, to cells, to tissues, to organs, to organ systems, to organisms, to populations, and even to the planet Earth as a whole.

Since the "mother theme" of this book is that all life fits together into one interwoven whole, that facts and examples are just little bits of one big, continuous, vast staircase, it's time we did a couple of things. First of all, we need to apologize for taking this universal whole of biology and immediately going against our mother theme by breaking biology down into separate chapters (We thought it would be easier for you to read and discuss if we presented the book in this way). Secondly, we would like to take this last brief chapter to demonstrate and re-emphasize our mother theme by pulling all the separate pieces back together once again.

If you recall, way back in the first chapter, we used the analogy of an escalator, one step moving into another to describe the flow of The Living Staircase. We also mentioned how, with time, you would begin to see how to incorporate new information into particular steps of The Staircase, and how all this information is interrelated. Well, this book should have served as a catalyst for this learning process. We hope you now understand how to take new biological information and, instead of memorizing it and storing it on a lone shelf in the basement of your mind, to integrate it into your own staircase of knowledge, to truly *understand* it. In other words, you should be able to fit any new knowledge into the larger context of the information you already have. This does not happen overnight. It takes a lot of practice.

Before we look at specific examples of how to do this, let's make sure we're on the same wavelength. Think of this exercise of fitting new information into context as mentally "walking The Living Staircase," carrying a new fact or idea, looking for its proper step (or steps), where it fits relative to the other information on that step, and then setting it down carefully. Then you should think about how the new information on that step alters what you know about the other steps nearby. Finally, think of The Staircase banister as the eight themes of this book. They help you move along the stairs, give you balance and support, and serve to connect the entire construction.

When presented with new information, one of your first goals is to identify where in The Staircase it fits. Next, think about the theme(s) on which the information touches. The hard part is when you then must walk The Staircase and place the information in the "right" spot. Not all information will connect with all levels and themes, nor will everyone put the same information in the same place, but you will be surprised by how interrelated things appear as you carry out this exercise.

Let's take the general example of **nerve signal transmission**.[1] There's no rule on where to start; you can begin on any step or with any part of the banister. Let's start with neurons, the specialized cells of nerve transmission. At first glance, it is clear that one part of nerve signal transmission, the neurons, belongs on the "cell step."

Now, how do these neurons relate to other levels of our staircase? If we think about the "molecules step" next, we'll recall that the molecules that are important for nerve signal transmission include at least: (1) those that make up the vesicular organelles that dump other molecules (the neurotransmitters) into the space between two neurons; (2) protein receptors on the "receiving neuron" that bind the neurotransmitter; (3) protein vacuum cleaners on the "sending neuron" that "vacuum up" extra neurotransmitter; and (4) the molecules in the receiving neuron necessary to pass on the signal, including ions to create the electrical gradient that is re-converted to a chemical signal as the message moves through the nervous system. The electrical gradient is created across the plasma membrane, another organelle, that is made up of proteins and lipids. The DNA in the nucleus encode these proteins, and metabolic processes in the cell produce the lipids. Enzymes, also proteins encoded in the DNA, carry out the metabolic processes. Cells, including the neurons with which we started, require energy for all their work and for their reproduction.

Okay, we're overdoing it here to make our point: everything is connected. But let's keep going. Neurons are part of the nervous system, which is essential for obtaining the food that drives our bodies. So we've walked to the "organ step" and the "organ system step." The nervous system is also composed of the brain, the spinal cord, and all the nerves connecting these and all parts of our body. Along with the other systems with which it is connected and (to use a human term) cooperates, such as the endocrine and digestive systems, the nervous system plays a role in sending the signals that (1) tell us we are hungry, (2) allow us to excite the muscles necessary to buy or hunt for food and prepare and eat it, and (3) stop us eating because we are full. The digestive system eventually creates constituents from this food that our bodies utilize to maintain nerve cells and create more if needed. As you can see, despite starting in different places on our staircase, we often return to the same point.

We could embellish these stairs for many more pages. We won't, but remember the nervous system is, of course, essential for many more processes than just eating.

---

1. As you have probably noticed, we directly refer to this example in Chapter 1 when discussing the steps of The Staircase, and in Chapters 2 and 5, as well as a related example, that of hormone signal transmission, in Chapters 7 and 8.

We don't know very much about the "nervous system step." It contains the mysterious brain, the workings of which we have only a crude understanding. Improper development or injury- or disease-induced malfunction of this system results in a range of symptoms, from discomfort, as in the case of a "pinched nerve," to death, as in the case of severe disorders of the central nervous system.

You can see how as you walk The Staircase, more questions arise and connections become clearer. Obviously, the molecules, cells, and organs of the nervous system affect the entire organism in which they reside. Thus, we've already walked close to our "organism step" while discussing the nervous system and the organs that compose it. It's almost impossible to discuss the one step without the other. Humans require the brain and its associated organs to perform any behavior—to eat, mate, talk, run, you name it.

The rest of the nerve signal transmission story in relation to The Living Staircase begins to fall into place: the nervous system (only for organisms that have one, of course) is the action/reaction center for interactive behaviors within and between populations—mating, cooperation, competition, violence, family behaviors. The information for potential behaviors is in the genes and the behavior is in response to the environment, but it is the nervous system that monitors and modulates the reception of a signal and the response to it.

We've taken a nice stroll through The Staircase carrying concrete examples and questions. Now, we get a little more abstract. We begin to study the banister of biological themes that runs alongside The Staircase. In other words, we try to blend the facts and observation that we were placing on The Staircase into the central themes. For example, interactions among neurons or displays of mating behavior are prime examples of one of these themes: communication. We've already discussed several examples of division of labor in nerve signal transmission: the varied molecules of the neuron performing different roles, the distinct roles of each neuronal organelle, and the different organs of the nervous system, each also having a distinct, though interconnected, job.

Homology and diversity, two other central themes, are two sides of the same coin. Ask yourself what similarities you know exist between nerve transmission in, say, mammals and insects. How are the nervous systems different and similar? Do both use the same neurotransmitters or are some similar and some unrelated? Is response to the neurotransmitters similar? Few people will know the answers to all these questions, but it is important that walking The Staircase, i.e. synthesizing what you learn, leads to new questions. The best way to enhance your staircase is to create more questions, then find out what answers are known; this will in turn lead to more questions.

Cycling and regulation are also often intimately related themes. For example, in the nerve transmission cycle, an organism must regulate its response to a nerve impulse. If there's a fire, and you start running, you need to stop running once you are safe; at some point, the excitatory neurotransmitter has to stop being released across, and must be removed from, the synaptic cleft. A regulatory recycling of neurotransmitter occurs when special proteins on the "signal-sending side" of the synapse re-absorb "unused" neurotransmitter for later use. You should be able to think of other examples of regulation and cycling related to neurotransmission that will strengthen your mental hike. Do not forget that classic regulatory mechanism: feedback inhibition.[2]

Nerve signal transmission is a pretty general example. Let's walk through a more specific example, selected randomly out of a textbook we read recently: "In culture, **cancer cells** can go on dividing indefinitely, if they have a continual supply of nutrients, and thus are said to be immortal." Imagine you are studying for a test in cell biology and you catch yourself memorizing, instead of learning, that last sentence. Usually, when we simply memorize information, it is because, without thinking about it, we have no pre-existing spot on The Staircase in our minds in which to place it. In one sense, as we mentioned earlier, "walking The Staircase" with that information is merely another way of saying "understanding" that information.

Okay, back to our mystery sentence. Parts of our staircase are built into the statement: "cells" are mentioned, and "indefinite division" should suggest a problem with regulation of the cell cycle. Review the cell cycle, its resting stage (interphase) and the other stages of the two types of cell division, mitosis and meiosis. Malfunctions in many different molecules, including those involved in both intra- and intercellular communication or in regulating the cell cycle can result in cancer (uncontrolled cell growth). Imagine a signaling process whereby a growth hormone moves from the pituitary gland to certain cells in the liver during early development. The signal tells the liver to keep growing. Molecules in the cell membrane bind and react to the hormone,[3] affecting changes inside the cell that signal the initiation of more protein production, leading to more cell parts, so that the cells can divide. The end result is the liver grows.

If at any point this process breaks down, for example if the feedback inhibition process that regulates the release of the growth hormone starts responding even without the presence of the hormone, the cell divides non-stop, causing cancer if the cell has no back-up regulatory system. Where else in this process could a problem lead to cancer?

---

2. See example of feedback inhibition in Chapter 5.
3. Chapters 2 and 7 expand the concept of receptors and cell signaling.

Continuing along The Staircase: cancer can affect any organ or organ system, since all of these organs and systems are composed of cells. Which organs would be most likely to become cancerous? Why? Obviously, cancer can affect the whole organism, as it can cause death and prolonged or severe pain. The disease has also had a great influence on the behavior of humans—as individual organisms and as a population. Not only have many scientists struggled to unravel the mysteries of the disease (research is, after all, a type of behavior), but as they have, many of us have shifted away from such habits as cigarette smoking and eating red meat. Scientists and doctors have shown that both these behaviors increase the chances of cells losing control of their growth, although we're not exactly sure how. A big part of cancer's place in The Staircase is the question: How much of cancer is hereditary and how much is due to the environment?

We clearly have a lot to learn about cancer and how and where it fits into The Staircase. Yet, beginning to walk with what we do know allows us to see what knowledge is missing. Then we can ask the important questions. Any disease that has the power to change behavior may also affect natural selection. (As you may have guessed, we are now beginning to use the theme-banister to enrich our understanding of cancer and its place in The Living Staircase.) Why do some human populations appear to be "immune" to certain types of cancers? Do some have a selective advantage? Why is cancer only now, historically speaking, taking such a heavy toll on humans? Was it not diagnosed before? Or is it only now, when people are living longer, that the disease "has a chance" to show its effects? Why haven't we evolved to the point where we can fend off cancer? Scientists are beginning to get hints of answers to some of these questions. For example, it appears that several different mutations may have to occur within the same cell in order for it to become cancerous. This suggests that we do have defense and back-up mechanisms to fight the disease.

We'll finish up the cancer example with some "homology banister," if you will—that is, how our knowledge of cancer fits into the theme of Chapter 6, homology. Mice, monkeys, and other model systems are very useful in studying cancer, its origins, and possible therapies against it. This is true because of the incredible similarity of the basic cellular processes that have been "saved" by evolution. Mitosis, meiosis and the molecules that are essential for carrying out and regulating these processes are very similar in mice and man.[4] Such homology means that studies of cancer and cancer therapies can first be carried out in mice or rats. Information gleaned from these studies does not guarantee success in humans, but it is a good start.

You're probably realizing by now how to walk The Living Staircase. With practice, you should be able to walk The Staircase with any piece of information that you learn in any biology class: development, cell biology, immunology, microbiol-

---

4. See Chapter 7 for a discussion of G-proteins, a specific class of proteins involved in this regulation.

ogy, genetics, ecology, botany, biophysics, anatomy and physiology, pathology, population genetics, or biochemistry. All of this vast knowledge is interrelated. Because you learn it in separate courses, in separate books, in separate chapters, it is your job to integrate. The connections, analogies, and examples we use in this book will help you to do just that.

Q & A with Jeff & Jen

JEN: Whew, well, I'm sort of glad we get the last word in all this.

JEFF: Why do you say that?

JEN: Well, this has been such a long and strange trip. It's all weird. Using street people and escalators to relate biology—I don't know. Maybe we can close this up with some straight biology.

JEFF: You mean something to memorize? Yeah, but didn't the book help you learn and understand how biology fits together? Couldn't you go out there now and walk that big Staircase?

JEN: Sure, but I guess I still don't really see the purpose in all this.

JEFF: It's just an analogy to force us to do something we often do subconsciously anyway. It's an exercise in understanding. We do it all the time in everyday life.

JEN: Like when?

JEFF: Like when we get any new information. If you find out you have a calculus test on next Thursday, you immediately fit that fact into your preexisting schedule. This means you have to set aside this much time to study on such and such a day, and that you'll have to finish your English paper a day earlier than you thought, and miss the play you wanted to see Wednesday night, etc. Your personal life and its schedule are another staircase.

We're constantly adding new information to our minds and shifting everything relevant to that information already in our minds to fit it.

JEN: Hmm . . . Okay, here's some "new information" for you—let's see what you can do with it: Your cat belongs to the phylum Chordata and the class Mammalia. Now, walk that Staircase.

JEFF: That's easy. First, don't make the mistake of thinking that all information has to fit neatly into every theme and every step of The Living Staircase. Just start with what you know.

Here goes. The cat is an organism. Scientists classify it based on two central biological themes: diversity and similarity. All animals with a backbone we classify as Chordata, and we classify Mammalia as organisms that have body hair, are born live, and have other characteristics. Based on this much of my walk, I would ask the following questions: What are the phyla called that do not have a backbone? What are the other classes within the Chordata phylum, and what makes them different from and similar to Mammalia?

I would then explore each of those similarities and differences a little bit. When did organisms evolve backbones? What are their advantages and disadvantages? How are other organ systems similar and different in a cat than in other members of Mammalia? I would expect cats' hunting senses, perhaps smell and night vision, to be exceptional. Thinking in terms of natural selection, why are some cats domestic and others not? When did cats become domesticated? What is the advantage to the cats of being domesticated? Can they revert to the wild or mate with wild cats? Why is . . .

JEN: Uh, excuse me, Jeff, I think I've got the idea. Let's go get some lunch.

Index

Transcription factor 89
Translation 87–88, 132
Tricarboxylic acid cycle 48
Trophic levels 174
Tuberculosis 40

V

Vascularization 166
Ventricles 53

Vesicles 191–193
Voles
    behavioral diversity 151
    monogamy in 153, 195
Volvox 27

W

Waggle dancing 32